WHALES
and Other Marine Mammals of the
EAST COAST

Tamara Eder
Ian Sheldon

Lone Pine Publishing International

© 2014 by Lone Pine Publishing International Inc.
First printed in 2014 10 9 8 7 6 5 4 3 2 1
Printed in China

All rights reserved. No part of this work covered by the copyrights hereon may be reproduced or used in any form or by any means—graphic, electronic or mechanical—without the prior written permission of the publisher, except for reviewers, who may quote brief passages. Any request for photocopying, recording, taping or storage on information retrieval systems of any part of this work shall be directed in writing to the publisher.

The Distributor: Lone Pine Publishing
1808 B Street, Suite 140
Auburn, WA USA 98001
Website: www.lonepinepublishing.com

Publisher's Cataloging-In-Publication Data
(Prepared by The Donohue Group, Inc.)

Eder, Tamara, 1974-
    Whales and other marine mammals of the East Coast / Tamara Eder.

    p. : col. ill., maps ; cm.

    Includes index.

    ISBN: 978-976-650-052-8

    1. Marine mammals—Atlantic Coast (U.S.) I. Title.

QL719.A85 E44 2012   599.5/091634

Illustrations: All illustrations are by Ian Sheldon except: Gervais's Beaked Whale (p. 126), True's Beaked Whale (p. 128), Gray Seal (p. 144), Harp Seal (p. 146) by George Penetrante; and Harbor Seal (pp. 138, 140), Northern River Otter (pp. 139, 148) by Gary Ross.

Photography: Ken Balcomb (p. 116); Renee DeMartin/West Stock (pp. 43, 64, 112, 142); Francois Gohier/ardea.com (p. 94); Tom and Pat Leeson/Image State (p. 76); NOAA/FLFWC (p. 68); photos.com (pp. 147, 150, 151, 154); Eric Stoops/Corel Corporation (pp. 50, 58, 90, 136); Kevin Walsh (p. 145).

PC: 27

Contents

Quick Reference Guide — 4

Introduction — 7

    Whale Origins and Evolution 8

    Behavior and Adaptations 10

    Extreme Whales 24

    Intelligence and Whale Research 28

    Whale Lore: from Legends to Modern Mythology. 31

    Whales in Peril 33

    Whale Watching 38

    About this Book 42

    Whale Tales & Record Breakers 44

Baleen Whales
page 46

    Rorquals 48

    Right Whales 66

Toothed Whales
page 70

    Ocean Dolphins 72

    Porpoises 114

    Beaked Whales 118

    Dwarf Sperm Whales 130

    Sperm Whales 134

Seals, Otters and Manatees
page 138

    Hair Seals 140

    Otters 148

    Manatees 152

Glossary — 156

Further Information — 159

Index — 165

About the Author — 168

About the Illustrator — 168

QUICK REFERENCE GUIDE

BALEEN WHALES

Northern Minke Whale • p. 48

Sei Whale • p. 52

Bryde's Whale • p. 54

Blue Whale • p. 56

Fin Whale • p. 60

Humpback Whale • p. 62

North Atlantic Right Whale • p. 66

TOOTHED WHALES

Rough-toothed Dolphin • p. 72

Bottlenose Dolphin • p. 74

Pantropical Spotted Dolphin • p. 78

QUICK REFERENCE GUIDE

TOOTHED WHALES

Clymene Dolphin • p. 80

Striped Dolphin • p. 82

Atlantic Spotted Dolphin • p. 84

Spinner Dolphin • p. 86

Short-beaked Saddleback Dolphin • p. 88

Atlantic White-sided Dolphin • p. 92

White-beaked Dolphin • p. 96

Risso's Dolphin • p. 98

Melon-headed Whale • p. 100

Pygmy Killer Whale • p. 102

False Killer Whale • p. 104

Short-finned Pilot Whale • p. 106

Long-finned Pilot Whale • p. 108

Killer Whale • p. 110

Harbor Porpoise • p. 114

QUICK REFERENCE GUIDE

TOOTHED WHALES

Cuvier's Beaked Whale • p. 118

North Atlantic Bottlenose Whale • p. 120

Sowerby's Beaked Whale • p. 122

Blainville's Beaked Whale • p. 124

Gervais's Beaked Whale • p. 126

True's Beaked Whale • p. 128

Pygmy Sperm Whale • p. 130

Dwarf Sperm Whale • p. 132

Sperm Whale • p. 134

SEALS, OTTERS AND MANATEES

Harbor Seal • p. 140

Gray Seal • p. 144

Harp Seal • p. 146

Northern River Otter • p. 148

West Indian Manatee • p. 152

Introduction

In 1969, we saw Earth from the moon. During the first moon landing, pictures of our planet were radioed back to Earth for everyone to see. These first images of Earth kindled a new kind of enlightenment and shifted our awareness of ourselves and our place on the planet. Seeing images of "Spaceship Earth" sparked two (among many) profound realizations that not only affected our thinking at the time but also would eventually affect the relationship between humans and whales. Primarily, the images reminded us that Earth is dominated by oceans. More than 70 percent of our planet's surface is covered by water, and the creatures that thrive in the marine habitat are marvelous in their form and adaptation. The second realization, rather less obvious, is that we may not be alone in this universe, and we conjured up a keen anticipation of finding non-human intelligence. One result of these combined realizations was a new appreciation for those animals that live in the sea, and when we looked into the eyes of whales and saw something familiar and intelligent, we redoubled our efforts to understand and protect these great creatures.

INTRODUCTION

Whales have fascinated people for thousands of years. At least 3500 years ago, the early Greeks used images of dolphins on mosaics, frescoes, vases and coins. Aristotle is believed to be the first person to propose that whales and dolphins are mammals, not fish, and he correctly identified and classified several different species.

When people speak of whales, they generally imagine the large whales that live in deep oceans. Broadly speaking, the distinction between whales, dolphins and porpoises is based on size rather than zoological affinity: whales are typically the largest of the three, and porpoises are the smallest. Some of the largest dolphins are larger than the smallest whales, however, so the terminology can be confusing. Technically, whales include the Gray Whale, rorquals, Bowhead Whale, right whales, Beluga, Narwhal, sperm whales and beaked whales. Dolphins include the Killer Whale, pilot whales, ocean dolphins and river dolphins. There are only six species of porpoises worldwide.

All whales, dolphins and porpoises are members of the order Cetacea, and they are commonly called "cetaceans." They are distinguished from other mammals by their nearly hairless bodies, paddle-like forelimbs, lack of hindlimbs, fusiform bodies and powerful tail flukes. At least 88 species of cetaceans are known to exist today, and they are scientifically classified into two suborders according to whether they have baleen or teeth. The baleen whales are in the suborder Mysticeti, and the toothed whales are in the suborder Odontoceti. There are only 11 species of baleen whales worldwide: the Gray Whale, rorqual whales, Bowhead Whale and right whales. The toothed whales number at least 70 species worldwide, including the porpoises, dolphins, sperm whales, Narwhal, Beluga and beaked whales.

Whale Origins and Evolution

The evolutionary history of whales is truly remarkable. Mammals originated on land, so cetaceans must have arisen from terrestrial ancestors. In fact, their closest living relatives are the artiodactyls: even-toed hoofed mammals such as cattle, antelopes, pigs and hippopotamuses. Artiodactyls and cetaceans have common ancestors that are unlike either of these present-day groups.

Mammals existed as early as 210 million years ago, but the great dinosaurs dominated the world until 65 million years ago, and mammals remained very small and insignificant. After the dinosaurs died out, mammals had the chance to diversify, and they evolved to walk the earth, fly in the sky and swim in the oceans. The first ancestor of whales known through fossil finds is likely the 50-million-year-old

Ambulocetus, an amphibious, dolphin-sized mammal that had hindlegs with webbed feet for swimming. Ambulocetus probably gave rise to Pakicetus, which still had legs, but had a more whale-like skull, teeth, jaw and tail stock. By the early Oligocene epoch (34 million years ago), several toothed whales existed, and Mammalodon, the baleen whale ancestor, also appeared. Once the basic physiology was established, cetaceans diversified dramatically, and by about 7 million years ago, the major forms of whales, dolphins and porpoises that we recognize today had evolved.

Over millions of years of evolution, the body plan of the whale changed dramatically from that of its four-legged progenitor to the oceanic form we see today. The loss of the hindlimbs and the modification of the forelimbs into flippers with no moving elbows are striking changes to the basic mammalian plan. Whales still retain a vestigial pelvis and femurs because the selective pressure for eliminating such structures is weak. The blowhole originated from the nostrils, which migrated to the top of the head and effectively separated breathing and eating. The ears were modified for hearing underwater, and echolocation developed in the toothed whales. The dorsal line of the whale adapted to create better propulsion underwater: the neck vertebrae stiffened, a dorsal fin originated, muscles and bones along the tail stock strengthened, and the tail flukes developed. The skull and jaws were highly modified, and then the division between baleen and toothed whales occurred. This division was paramount because the baleen whales became specialized for catching large quantities of small creatures, whereas the toothed whales could catch smaller quantities of larger prey. Other modified features common to all whales include a highly streamlined body and a nearly total loss of body hair.

Pakicetus

Although whales and dolphins evolved in the oceans, some species now live in fresh water. The river dolphins, which number five species worldwide, inhabit large river systems such as the Amazon, Orinoco, Indus, Ganges and Yangtze. As well, most porpoises can travel freely between fresh and salt water,

INTRODUCTION

as can the Beluga (which does not live in our area). Generally speaking, however, the saltwater whales tire quickly in fresh water because of their reduced buoyancy, and their skin might wrinkle and slough (just like ours does in a bath).

The fact that cetaceans look a bit like sharks is a tribute to convergent evolution. Sharks and cetaceans are not closely related at all, and there are many physiological differences between them. However, a gray-colored shark and a dolphin, for example, may be difficult to distinguish at first glance. Their similar appearance reflects a desirable body plan for being a successful ocean creature.

Although mammals have many characteristics that differentiate them from other animals, the most obvious differences between whales and sharks are that whales breathe air and nurse their young, whereas sharks have gills and do not nurse their young. The easiest way to distinguish between sharks and whales, however, is to look at their tails: the flukes of a cetacean are horizontal; the tail of a shark is vertical. Many frightened swimmers have thought that an approaching dolphin was a shark, but a shark swimming close to the surface of the water reveals two triangular "fins" above the surface: its dorsal fin and the top of its tail. Dolphins reveal only their dorsal fin.

Shark

Dolphin

Behavior and Adaptations

Feeding with Baleen

The baleen whales, or mysticetes, are typically the largest of all cetaceans. Of these, the largest are the females. Ironically, these large whales feed on some of the tiniest ocean creatures.

Baleen plates are a remarkable new mammalian adaptation that permits efficient filter-feeding of plankton, tiny crustaceans, small fish and some mollusks from the ocean water. Baleen was once called "whalebone," but it is really not

INTRODUCTION

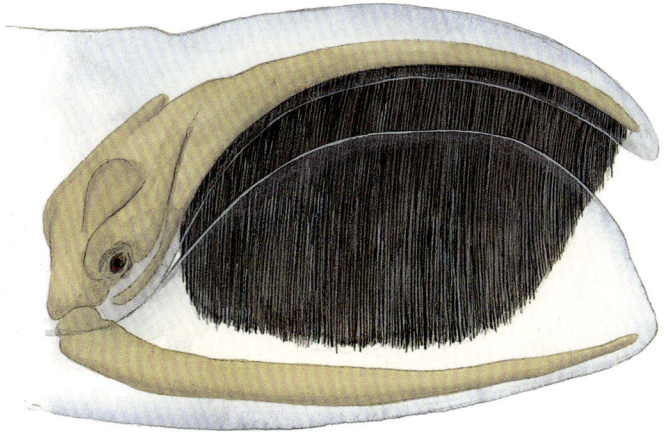

Baleen specialized for skim-feeding (Bowhead Whale)

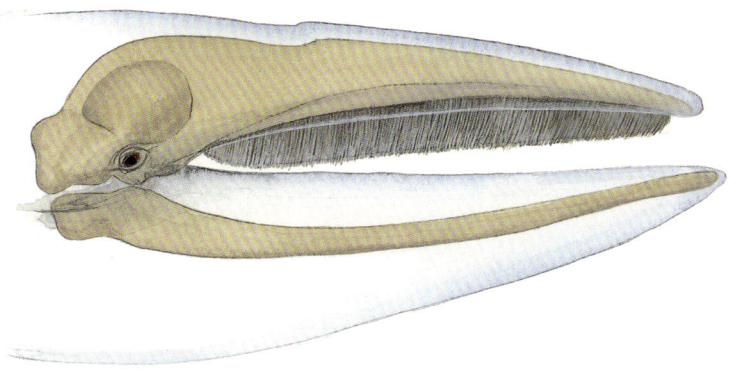

Baleen specialized for lunge-feeding (Northern Minke Whale)

bone at all. Although baleen grows where teeth normally would on other mammals, it is actually made of an entirely different keratinous material very similar to human fingernails and hair. The baleen plates grow down from the gum of the upper jaw, and they are shaped like very long, narrow triangles. All the plates hang one against the other, similar to vertical blinds on a living room window. The hairs on the edge of each plate face into the mouth.

Lunge-feeding (sometimes known as gulp-feeding), skim-feeding and bottom-feeding are the three basic styles of feeding among baleen whales. Within each style, there is variation between species and even individuals.

INTRODUCTION

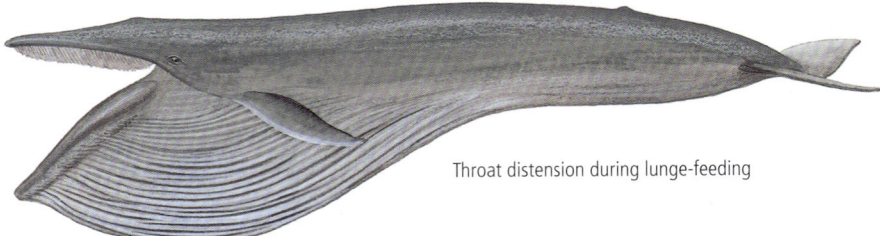

Throat distension during lunge-feeding

The rorqual whales primarily use lunge-feeding. If necessary, prior to surface lunging, the whale herds its prey into a tight group near the surface. Once the prey is concentrated, it lunges and opens its mouth to gulp in tons of food-rich water. A rorqual has specialized throat pleats that expand like an accordion and allow the throat to distend in a balloon-like way. These pleats contract automatically as the whale closes its gaping mouth. The mouth is not closed completely, but just enough to permit the baleen to meet the lower jaw and create a sieve. The water is squeezed out through the baleen, while the krill, copepods and other small creatures are trapped inside the mouth against the baleen. Rorquals have short baleen that is continuous around the front of the jaw, making the mouth an effective filter when it is nearly closed.

Humpback Whale bubble-netting

INTRODUCTION

A famous variation of lunge-feeding is bubble-netting, used by Humpback Whales. One or several Humpbacks begin releasing bubbles as they swim in a circle below a school of fish. As the bubbles rise, they create a curtain that disorients the fish and traps them in a central area. Once the fish are concentrated inside this bubble-net, the whales surge up in the middle of the "net" and gulp the creatures into their gaping mouths.

Skim-feeding is passive in comparison to lunge-feeding, and it is a style used primarily by right whales. The Sei Whale, though it is a rorqual, also frequently uses skim-feeding. For a long time, skim-feeding was believed to occur only at or near the surface, simply because we could only see the whales doing it there. Skim-feeding often occurs at great depths, however, and only recently was this activity recorded on film by researchers. This feeding style requires the whale to swim through a school of plankton or other small creatures with its mouth open.

Right whales do not have baleen in the front of their mouths, and the baleen on the sides is very long. When the whale opens its mouth, the opening is like a small cave with walls of baleen. As the whale swims through the food-rich water, the water enters the mouth and continues right out through the baleen. Food also goes into the mouth, but it is trapped by the long baleen. The throat of these whales does not extend, and the mouth is kept still and open as the whale swims. Once many food creatures have accumulated in the mouth, the whale closes its jaws and swallows.

The Gray Whale (*Eschrichtius robustus*), found only on the west coast, is the only bottom-feeding baleen whale. It feeds by scraping up bottom sediment and filtering out the invertebrates.

North Atlantic Right Whale skim-feeding

Feeding with Teeth

The toothed whales feed on creatures larger than those fed on by baleen whales. Their jaws are modified according to their style of feeding and their prey. Dolphins, for example, often have long, narrow jaws that are fast-closing and capable of catching small, darting fish. Killer Whales have wider jaws with heavy, conical teeth for holding prey and tearing off large chunks. The size of their prey matters little to Killer Whales, and certain groups of these predators are known to attack other cetaceans, such as porpoises, dolphins and even larger whales and especially their calves. Porpoises have spade-shaped teeth that are flattened side to side. These teeth are effective for slicing and shearing rather than holding prey.

The Sperm Whale and the beaked whales are believed to be suction feeders—by quickly moving the tongue backward in the mouth, they create a powerful suction to pull in their prey—and there is much debate about how useful teeth are to them. The Sperm Whale has a very narrow jaw, relative to the size of its head, with heavy, conical teeth. The white mouth of the Sperm Whale may even act as a lure to bring its prey close. One adult Sperm Whale that was killed had no lower jaw at all—an astonishing injury that appeared to have had little effect on the overall health of the whale, which was large and clearly well fed.

Beaked whales have some of the strangest teeth and jaws of all the odontocetes. Usually, only the male's teeth erupt, and the females feed without any functional teeth. Males have two (or sometimes four) teeth in the lower jaw, either close to the tip of the beak or in the middle. The teeth tend to be large and triangular, and sometimes they overlap the upper jaw on the outside of the mouth.

Killer Whale teeth

Dolphin teeth

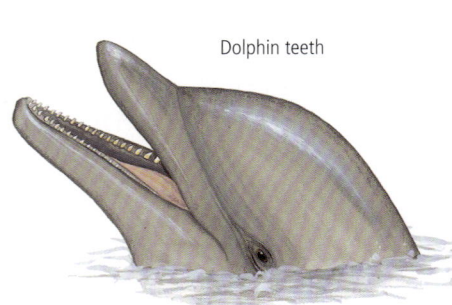

INTRODUCTION

The Blowhole

One of the marvels of cetacean anatomy is the blowhole. Unable to breathe through its mouth, a whale inhales and exhales exclusively through its blowhole. The blowhole is designed for rapid air exchange, and it is the result of an evolutionary change in skull shape and a slow migration of the nostrils and breathing passage to the top of the head—away from the throat. In some cases, most notably the Humpback Whale, air can voluntarily be released from the lungs into the mouth if the trachea is slightly dislodged, but this action does not give the whale the ability to breathe through its mouth.

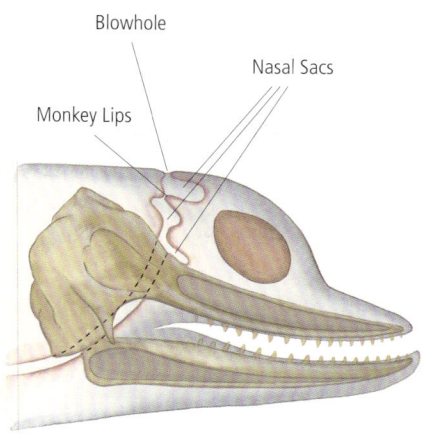

For whales, breathing is voluntary, and they open and close their blowholes at will. The baleen whales have two blowholes that are controlled by strong muscles attached to the upper jaw. When the whales surface, they make an explosive blow and inhale in such quick succession that the entire process takes only a few seconds. The toothed whales have only one external blowhole. In fact, they have two nasal passages, each with the ability to open and close, but these passages (called "monkey lips") are hidden internally. The single external blowhole is a new adaptation unique to odontocetes that allows for sound generation by the hidden soft tissues of the nasal passages.

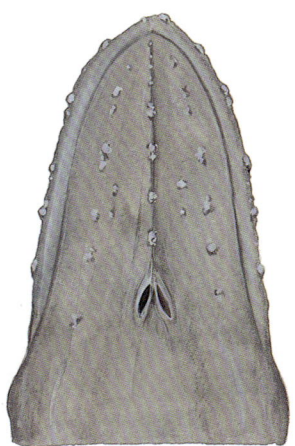

Paired blowholes of baleen whales

Single blowhole of toothed whales

INTRODUCTION

Sleep

Unlike terrestrial mammals, cetaceans do not enter a deep state of sleep. They must rest, but they require much less sleep than terrestrial mammals. Evidence suggests that whales require less sleep because the marine environment induces a brain state that resembles sleep. Studies show that even humans, when immersed in water for up to six days, may require only one hour of sleep a day instead of the normal eight hours. Moreover, whales are voluntary breathers, so for them to breathe at the surface, to keep moving to prevent stranding and to watch for potential dangers, they must never fall soundly asleep.

To relax, many whales use the technique of "logging" at the surface. Logging is a form of rest where whales float almost motionless at the surface, and if there are two or more, they will all point in the same direction. Their breathing and heart rates are lowered, and their eyes may close, but they have not totally lost consciousness. Mariners once believed that a logging whale was fast asleep and that even an approaching boat could not wake it, but they were mistaken. Dolphins and porpoises are able to "sleep" half their brain at a time.

Lungs and Diving

Perhaps the greatest danger for a deep-diving air breather is nitrogen narcosis, a desensitized state caused by pressurized air entering the circulation system. The other main hazard, the bends, occurs when bubbles spontaneously form in joints and tissues as a result of rising to the surface too fast. Whales are not affected by either of these conditions because of several adaptations that evolved over time.

Before a deep dive, whales remove the air from their lungs. Air is highly compressible, whereas water is not. Like our bodies, most of a whale's body is water and can therefore handle the pressure of a dive. Only the air in the lungs is subject to compression. Any air that remains in a whale's respiratory system is pushed into the inflexible windpipe, where it cannot be compressed. This strategy prevents air compression and thereby effectively

Sperm Whale diving

eliminates nitrogen narcosis. The bends, also known as decompression sickness, is prevented by the efficient and rapid transport of nitrogen back into the lungs as the whale surfaces. Circulation of blood through the muscles during the dive is also reduced, which further reduces the possibility of nitrogen bubbles forming.

Sounds and Songs

In water, sound travels almost five times faster than in air. A cetacean uses sound to keep in touch with other members of its group, and some of the odontocetes use sound for echolocation. Different whales make different sounds, and if the sounds have meaning, one species likely cannot understand another species. Sounds vary from loud to quiet, and from deep rumbles and belches to high squeaks and clicks.

Baleen whales commonly produce the deepest and loudest sounds. The vocalizations of a Blue Whale can be less than 20 hertz in frequency and up to 188 decibels in volume. Sophisticated military technology capable of detecting submarine sounds from far away has been able to pick up the sounds of a Blue Whale from over 1500 miles away. This kind of evidence changes how we look at a group of whales. Perhaps whales do not have to be in visual range to be considered part of the same group. Two rorqual whales seen 30 miles apart may not be "solo" whales.

Humpback Whales make extremely loud and long songs. Samples of the unique and complex songs of Humpbacks have even been sent out on *Voyager*s 1 and 2, spacecraft that are on their way toward other worlds in this galaxy. The Humpback song samples are in a section of recording that includes greetings in 55 languages from 60 different countries.

Although Humpbacks are now part of the ambassadorial elite of Earth, no one actually knows how they and other baleen whales produce sound. Baleen whales have a larynx, but they have no vocal cords, and their paired blowholes release no bubbles when they sing. A laryngeal source for the sound is not ruled out, but much more research must be done to determine how mysticete whales make sound.

Blue Whale

In odontocetes, sound production is complex, but it is better understood than sound production in mysticetes. The external blowhole of the toothed whale hides the "monkey lips," or the internal soft tissue of the dual inner nasal passages. The monkey lips can pass air to generate sound, much like we can purse our lips to making squeaking and kissing sounds. One monkey lip can be open while the other is closed, creating a variety of different noises. All the while, the external blowhole stays shut, so no air is released. The sounds made by odontocetes are primarily whistles, squeaks and clicks emitted in a variety of patterns.

Many groups of toothed whales have unique "dialects" different from those of other groups of the same species. All Killer Whales, for example, produce many similar sounds that are probably understandable from one whale to the next, but closer analysis reveals that one group will produce sound combinations that no other group does. Dolphins also have very sophisticated sounds and sound patterns. Many species have "signature whistles" for each individual, which are probably like human names. Signature whistles are often accompanied by a small string of bubbles released from the blowhole.

Echolocation

The use of echolocation, though it is rare in the animal kingdom, is not unique to whales. Bats, shrews and certain birds are efficient echolocators, and there may be other animals that we are not aware of yet that also use echolocation. Echolocation involves sending out high-frequency pulses or clicks and listening to how these sounds bounce off objects present in the vicinity. Echolocation, particularly the underwater variety, is also known as sonar (SOund Navigation And Ranging). Military organizations once applauded themselves for discovering and using such "advanced" technology. Some of this military pride was shattered upon learning that bats and whales had been doing it for millennia.

As far as we know, only the odontocetes use echolocation. When toothed whales emit the echolocation clicks, the sounds are passed through the "melon"—a specialized structure in their forehead that focuses the

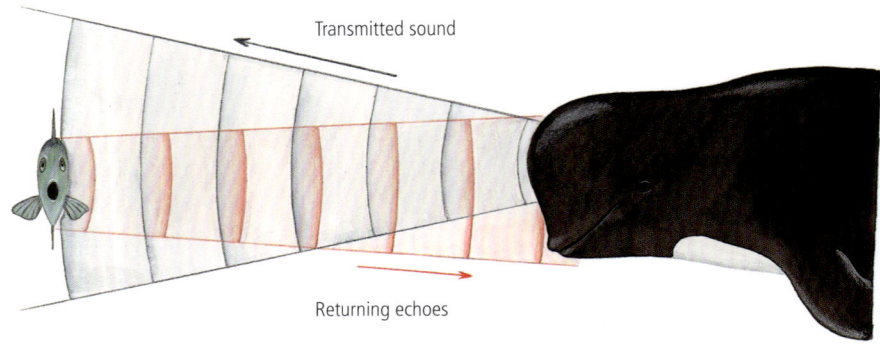

Transmitted sound

Returning echoes

INTRODUCTION

sound. The melon is mainly low-density oil, and it gives toothed whales their distinctive, bulbous forehead. When the outgoing signals hit an object, they bounce back in an altered form. These returning "echoes" are so detailed that the whale can perceive shape, size, texture, distance and probably many more characteristics. The returning signals are received through the lower jaw, which helps transmit the sound to the modified ear.

Echolocation is not limited to surfaces. Evidence indicates that odontocetes can "see" inside things, even inside other bodies, to learn which females (human, dolphin or otherwise) are pregnant. Many interesting research projects involve dolphins and pregnant women; the dolphins seem to be aware of the pregnancy and act differently around these women.

The primary use for echolocation in whales, however, is for finding food and determining their immediate surroundings. The pulses are effective for at least 2500 feet, and they allow the whales to hunt even in totally dark or murky water. Some species, such as the Sperm Whale and even the Pantropical Spotted Dolphin, may use such intense pulses that their prey is momentarily stunned, making feeding much easier.

Migration

Both mysticete and odontocete whales may perform migrations of some sort. The baleen whales typically make long migrations between their feeding waters and their calving and mating waters. Most toothed whales are more appropriately called nomadic; their journeys follow food cycles rather than predetermined paths related to certain activities. Sperm Whales, bottlenose whales and Long-finned Pilot Whales make seasonal migrations, but very little is known about the migrations of toothed whales.

In general, and in both hemispheres, baleen whales migrate from high-latitude feeding waters to low-latitude calving and mating waters. The whales often feed only in the high-latitude waters, and during periods of calving, mating and migrating, they consume little or no food. The foods that baleen whales feed on—primarily plankton, krill, copepods and small fish—are particularly abundant in the high-latitude waters. Some theories suggest the baleen whales migrate to warm waters each year because their calves may not be able to survive in the cold though food-rich water, but there is little evidence to support such claims.

The details of why whales migrate may forever remain a mystery to us, as may the mechanics of how they unerringly traverse thousands of miles of water. The long journey of migrating baleen whales takes two or three months in each direction. The whales travel nearly constantly at speeds of just a few miles an hour. The cues whales use to navigate are not well understood. Perhaps they spy-hop occasionally to follow natural coastal features, or they may be able to detect subtle changes in water "tastes" and currents. Some researchers suggest that whales can use

INTRODUCTION

the sun and stars as a compass like we do, or they may be able to follow sea floor topography. Some cetaceans have the mineral magnetite (iron oxide) present in their bodies, and this is a key substance for biomagnetism (biomagnetism is when an animal can orient itself to the magnetic field of the earth, and perhaps use this ability for traveling long distances). Whichever way they do it, whales continue to dazzle the researchers with their ability to travel and properly navigate over long distances.

Strandings

Strandings are staggering events where whales end up in shallow water and can no longer maneuver. If the tide goes out, the whale is left dry on the sand, where it usually dies. Stranding can happen to one whale or to hundreds of whales at a time.

No one seems to know why some cetaceans become stranded. Some people think that in areas of gently sloping beaches, the echolocation of the odontocete whales is distorted. It is true that strandings occur more often with deep-water species, such as Sperm Whales or pilot whales, as if they become disoriented if they get too close to the coastline. Cetaceans that regularly live in coastal waters, such as dolphins and porpoises, rarely strand. In these cases, the individuals that strand are usually injured or diseased. Belugas, which are found north of our region, may strand when high numbers of them congregate in estuaries and shallow bays (a stranded Beluga will live until the next tide comes to free it as long as a polar bear does not find it). Some Killer Whales, notably those on Argentina's eastern coast, intentionally strand on rocky beaches where colonies of sea lions are near the water's edge. The Killer Whales can then grab sea lion pups and wiggle their way back into the water. Likewise, some Bottlenose Dolphins work together to push schools of small fish up onto muddy banks, where the dolphins can easily grab them and then slide back into the water. Other than these examples of intentional stranding, a stranded whale is one whose life is at serious risk.

If you find a whale alive and stranded, do not attempt to touch it or help it. In some countries, it is illegal to help whales unless you have proper training. Unskilled "help" for a stranded whale has often resulted in greater harm to the whale and injuries to the person. The first thing to do in the event of a stranding is to contact the nearest wildlife authorities or the police. If necessary, the immediate first aid for a stranded whale is to make sure its blowhole is unobstructed by sand or debris so it can breathe, and to keep its skin moist so it does not get sunburned or dehydrated.

Long-finned Pilot Whale

Displays and Surface Behavior

BREACHING: This behavior is a favorite for whale watchers. A breach is when some or all of the whale's body rises out of the water and splashes back in. Why whales breach is not clear. They may enjoy it, it may help rid the body of barnacles and parasites, or they may do it if they feel insecure or threatened. In some cases, whales breach repeatedly, much to the delight and astonishment of nearby whale watchers. Although many people assume that such instances are positive for both the people and the whale, some researchers suspect that the whale is breaching because of distress from boat harassment. Dolphins may breach in a high leap clear out of the water and spin or somersault while in the air.

LOB-TAILING: A lob-tail is when a whale forcefully slaps its flukes on the surface of the water while its upper body remains submerged. Some species are known to lob-tail as a sign of aggression, whereas others appear to do it for much more benign reasons, such as communication.

SPY-HOPPING: Most whales spy-hop, especially while socializing. Spy-hopping is when a whale raises its head almost vertically out of the water as if to get a better look around. This may be exactly what it is doing because the whale typically rises only until its eyes are exposed and no more.

FLIPPER-SLAPPING: Cetaceans may occasionally roll onto one side and slap their flippers against the surface of the water. This behavior may be repeated several times in a row, but we do not understand what it signifies.

INTRODUCTION

FLUKING: Before a whale dives, it frequently raises its flukes above the surface of the water. For a lot of species, this action indicates that a long, deep dive has started. Many researchers use the markings and scars on the flukes of whales to identify and catalog individuals.

LOGGING: For most species, one to several whales may "log" at the surface as a form of rest. When several individuals log, they all face the same direction and stay in a close group. Logging requires minimal effort, but it is not an unconscious state because whales must remain at least semi-conscious at all times in order to breathe.

BOW-RIDING: Dolphins, porpoises and some small whales frequently engage in bow-riding and wake-riding. Dolphins appear to enjoy this activity, and people certainly enjoy watching them. The pressure wave created by a moving boat pushes the dolphin along at great speed with minimal energy output by the animal.

PORPOISING: Dolphins and some small whales may break the surface of the water in a low-arcing leap each time they take a breath. This fast mode of travel is called "porpoising," and it is very energy efficient. Air has less resistance than water, and the momentum the cetacean built up underwater propels it a great distance during the leap. In spite of the name, this behavior is not common among true porpoises.

RUBBING: Rubbing is a behavior done by several species and is probably to massage the skin or to aid in molting—or simply to feel good. In some regions, Killer Whales massage themselves on shallow underwater beaches of rounded pebbles, an activity that is sometimes visible to a boater above.

PLAYING: Probably all cetaceans engage in playing activity. Play can be anything and may involve juveniles learning particular behaviors that will aid them as adults. If you see whales or dolphins playing at the surface, enjoy watching them, but try not to disturb their fun or harass them.

INTRODUCTION

BLOWING: One of the first indications that a whale is nearby is the distinct blow it makes upon surfacing. The blow of a whale is visible because of a possible combination of factors. The difference in temperature between the whale's lungs and the outside air, some seawater from the surface of the animal's skin and accumulated droplets of mucus and oils from the whale's lungs and nasal passages all contribute to the blow. The shape and size of the blow varies from species to species. Within a species, blow shape and size also vary. Some species, such as the Sperm Whale, have such unique blows that seeing the blow alone is sufficient to identify the whale. Typically, the first blow after a long dive is louder and bigger than succeeding blows. Blows may be diagnostic of the species when viewed from the front or rear. Side views may not be as distinct. The blows illustrated below and on the following page are to scale and are all as viewed from the front.

DIVE SEQUENCE: The most frequently observed activity at the surface is the dive sequence. Beginning with a blow, the dive sequence reveals certain characteristics about the animal. The shape of the head and blowhole, the length of the back, the dorsal fin, any dorsal knuckles or bumps, the tail stock and the flukes are all features that may be seen in a dive and that can be diagnostic enough to identify the species.

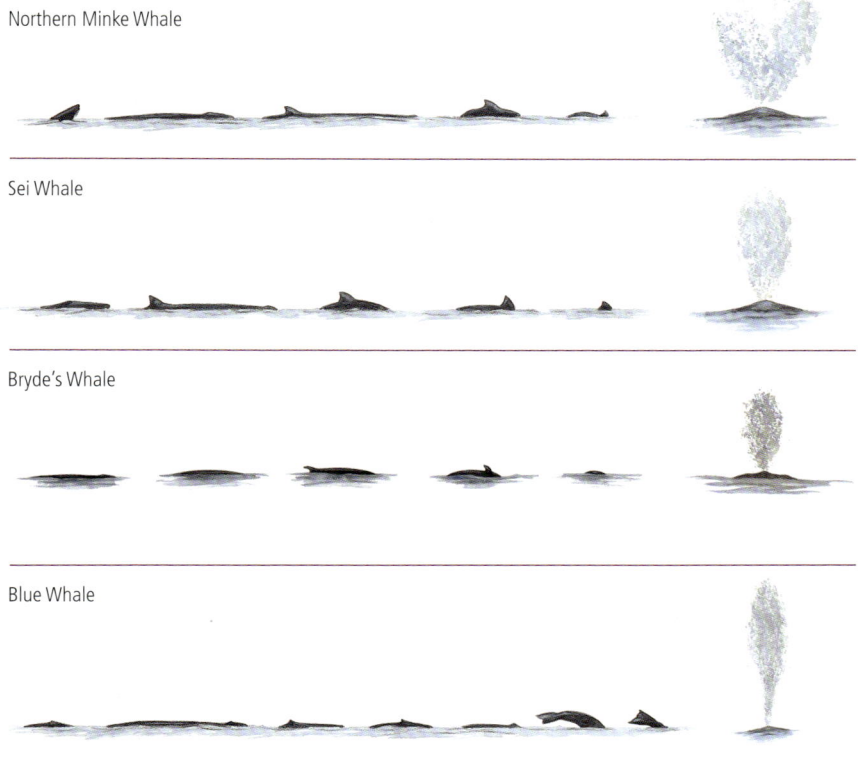

Northern Minke Whale

Sei Whale

Bryde's Whale

Blue Whale

INTRODUCTION

Fin Whale

Humpback Whale

North Atlantic Right Whale

Sperm Whale

Killer Whale

Extreme Whales

A simple truth about cetaceans is that many of their characteristics are difficult for us to understand. Their size, their weight, their communication—much of what makes a whale a whale—is unlike anything else we know. Typically, our efforts to understand whales involve compartmentalizing the information so that it is more manageable for us. We like to discuss large whales in terms of minimum and maximum lengths and weights, and we frequently discount reports of overly large whales as either myth or inaccurate reporting.

The framework in which we view whales, however, is not absolute. If the maximum length of a Blue Whale is reported by several authorities to be 100 feet, that does

INTRODUCTION

not preclude a whale of 105 feet, or more, from existing. In fact, there is much evidence to support the idea that large whales may keep getting slowly larger the older they get. Whalers of 100 and 200 years ago recorded enormous lengths for many whale species. Because whalers typically took the biggest whales (a simple choice to maximize profit for energy output), they would have taken the oldest whales. Researchers commonly suggest that the reason we do not see such large specimens now is that the whales currently alive are comparatively young. Such a suggestion has to be based on the assumption that whales keep growing. Can this assumption be proven scientifically?

Length and Weight

Few animals demonstrate extended growth, but all that do share similar characteristics in their bone structure. For an animal to keep growing in length, the vertebral column must also be able to continue growing. Animals that stop growing soon after sexual maturity, such as humans, have vertebral columns that cannot grow anymore. On either side of each vertebra there is a cap-like bone called the epiphysis. While the animal is growing, these "caps" are separated from the vertebra by a cartilaginous plate. For most mammals, after sexual maturity the caps fuse to the vertebra, and growing (in length) stops. For large whales, such as the rorquals, this fusion is delayed well past sexual maturity, and many older whales that have been caught have been found to have unfused epiphyses.

Given this evidence, it should not be surprising to find accounts of extremely large whales. For small whales, dolphins and porpoises, the epiphyses probably do fuse comparatively early, and the animals do have maximum lengths. In most cases, the measurements in this book indicate both the maximum records for each species and the average size. This average is the most likely size of whale for each species that you may see while out whale watching. A truth about nature is that there are always exceptions. For example, the maximum length given in this book for the Humpback Whale is 62 feet. This figure represents a reasonable maximum for the species. However, one female Humpback captured in shallow waters near

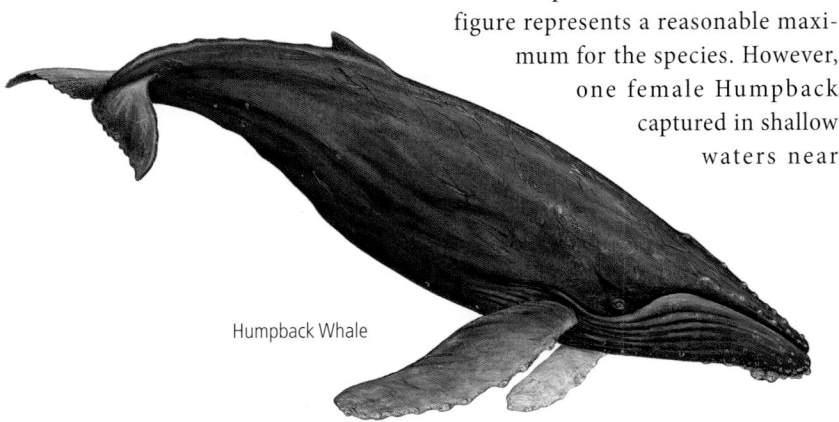

Humpback Whale

INTRODUCTION

Bermuda measured a staggering 88 feet long. Exceptional whales do exist, and they remind us that our measurements and descriptions are guides, not rules.

The weights of the large rorqual, Bowhead and right whales are all subject to revision as we discover greater or more exceptional examples. Their weights are difficult to determine, but we can reasonably assume that the larger a whale specimen is, the more it weighs. The maximum weights recorded in this book are from unusually large whales. The generally accepted average weight is also included, and this figure corresponds to the average length.

Diving Depth

Other whale extremes are also well documented, such as the depth to which Sperm Whales can dive. Good science is about taking new information and shifting our framework to encompass it, rather than discounting evidence as mere calculation error or myth.

Sperm Whales were once thought to dive to a maximum depth of about 5000 feet. We now accept that Sperm Whales can dive much deeper. One of the deepest records is of two male Sperm Whales that were killed upon surfacing from water that was at least 10,000 feet deep. Both of them had freshly eaten, bottom-dwelling sharks in their stomachs, indicating that Sperm Whales can dive as deep as the sharks lived. Perhaps there is no maximum; the Sperm Whale is perfectly adapted to handle the stress of deep diving, and the determining factor of the dive depth may be availability of food or time without oxygen.

Longevity

Current research on Bowhead Whales has turned up startling evidence that these Arctic-dwelling whales may live for more than 200 years. Interest in Bowheads sparked when Inupiat hunters found old spear points and arrowheads buried deep in the flesh of a few individuals. Some of these stone points are dated at up to 200 years old, indicating the whale could be several years older than that. The traditional methods used by researchers to determine the age of a whale do not work on Bowhead Whales. A new method that studies the delicate layering in the lens of the eye is being employed to determine Bowhead ages. This method has shown several individuals at 150 years or more, and at least one at

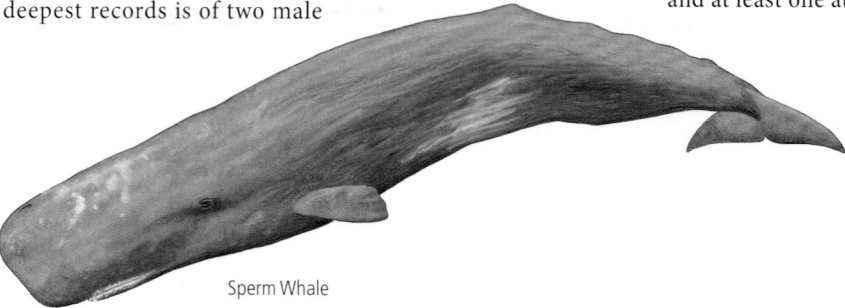

Sperm Whale

INTRODUCTION

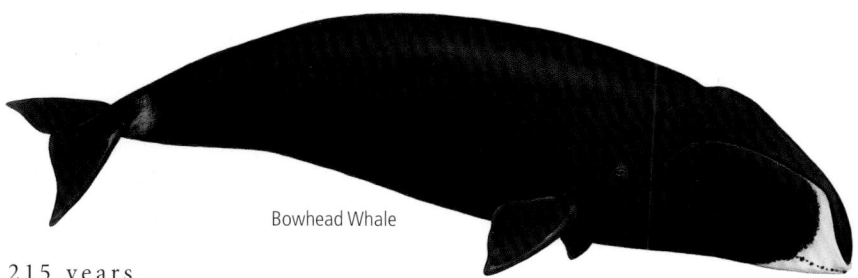

Bowhead Whale

215 years old. We do not know the maximum life span of a Bowhead Whale, and some researchers suggest that 215 may be conservative. At more than 200 years old, Bowhead Whales are now considered to be some of the longest-lived creatures on Earth, together with giant tortoises (150 years old, with disputed records at up to 200 years old) and giant clams (up to 220 years old). The Blue Whale is also considered a long-lived species, but we have yet to determine the age extremes for this species.

Whales vs. Dinosaurs

It is a popular belief that whales are bigger than dinosaurs. Although whales can be accurately measured for length, weighing the largest of these beasts is nearly impossible. As for dinosaurs, measuring their length involves speculation and educated guessing. Many dinosaur species are known only from a few separate bones, and determining the length of a complete animal is sheer inference. To complicate matters, paleontologists keep turning up larger and larger specimens. No one even knows which species is the biggest dinosaur. Weighing a dinosaur is, of course, impossible, and any given measurement is only theory.

Certain rules in biology make this guessing game easier. We know, for example, that a marine animal can generally have a greater body mass than a land animal because the water helps support its weight. The conclusion is that Blue Whales—at up to 200 tons—win as the heaviest creatures, whereas certain dinosaurs are the longest creatures that ever lived. The longest dinosaurs known include the Supersaurus, at 100–130 feet, and both the Seismosaurus and the recently discovered Argentinosaurus, at 110–160 feet.

INTRODUCTION

Intelligence and Whale Research

A common question people ask is, "How intelligent are whales?" The honest answer is that we do not know. Many studies have been done with whales to determine their intellectual abilities, but we do not have accurate ways of measuring intelligence in animals other than humans. Also, whales, dolphins and porpoises differ in intelligence amongst themselves. For example, most people accept that Killer Whales and Bottlenose Dolphins are more intelligent than river dolphins. Even how we define intelligence is ambiguous: it can mean independent thinking, analytical skills, understanding and evaluation skills or the ability to use reason and judgment in daily life. Intelligence can also be measured as a correlation between brain size and brain complexity.

Another difficulty arises in how we test cetacean intelligence. We require whales and dolphins to learn our signals and sounds, and then complete performance and ability tests. The inherent flaw in these procedures is that we are not measuring how intelligent a whale is with such tests, we are only learning how well it does on tests designed by us. By learning the sounds, signals and performance abilities of whales and dolphins in their own environment, we may learn far more about the intelligence of these creatures.

Looking at the brain of a cetacean can give us an indication of its potential intelligence. Unfortunately, the ratio of body size to brain size (also called the "encephalization quotient") does not directly correlate with intelligence level. In absolute terms, Bottlenose Dolphins have larger brains than do humans, but their encephalization quotient is smaller. Bottlenose Dolphins and chimpanzees have a similar encephalization quotient. The complexity of a Bottlenose Dolphin's brain is higher, however, and it has a well-developed cerebral cortex. (The cerebral cortex is known to control the coordination of voluntary movement with

Bottlenose Dolphin

INTRODUCTION

the senses, but it also contains centers responsible for memory, abstract thought, language and intellect.) Some people argue that this level of complexity serves to coordinate this dolphin's motor processes and echolocation, rather than being a sign of true intelligence. However, both odontocetes and mysticetes show a high level of complexity (relative to other mammals), and because mysticetes are not known to use echolocation, there is no clear evidence to support that argument.

Problem solving and the ability to learn are two well-accepted ways to assess intelligence. Whales in the wild show signs of teaching each other new things they have learned, and certainly mothers teach their calves many techniques for living. In captivity, cetaceans learn complex routines very quickly, and they even learn our language symbols. Studies with Bottlenose Dolphins indicate that they understand some abstract concepts like "sameness" and "difference." Using sign language, a dolphin easily understands the difference between "take the ball to the hoop" and "take the hoop to the ball." Dolphins seem quite capable of learning the basics of our language and signals, but we still have a long way to go to learn their sounds and language. Researchers debate about definitions of language, and whether or not cetaceans have true language. Nevertheless, it is clear to any observer that dolphins use sounds between one another, and these sounds appear to have intention and even meaning.

Some dolphins and whales appear to have good levels of problem-solving ability, and there are several well-known examples of problem solving in wild cetaceans. In the wild, Bottlenose Dolphins have been observed using tools in problem situations, such as wielding the spines from a dead poisonous fish to oust an eel from its crevice. Many researchers believe that the cooperative bubble-netting demonstrated by Humpback Whales is an intentional act of cooperation to solve a common problem, although other researchers think this behavior may be simply instinctive. During the heavy whaling years two centuries ago, Sperm Whales were frequently reported to swim and dive in directions upwind when pursued by a sailing vessel. Because sailing vessels cannot travel directly upwind, this behavior could indicate a high level of reasoning or problem-solving ability in Sperm Whales, as well as an awareness of the consequences of being caught.

Problem-solving and behavioral studies are excellent forms of research that continually lead to a greater understanding of cetacean ability. The difficulty with problem-solving studies is in differentiating between instinctual responses and actual judgment and evaluation.

In the United States and neighboring regions, whale research in the wild is becoming much more sophisticated. Satellite transmitters are being attached to many species to better understand their migration patterns. Photographic catalogs

INTRODUCTION

are used to identify hundreds of individual whales of many different species, especially Killer, Humpback and Blue whales. Killer Whales are reliably identified by the shape and scarring of their dorsal fin and by the shape and coloration of the "saddle" patch directly behind the dorsal fin. Humpback Whales can be recognized by the undersides of their tail flukes. The fluke coloration, shape and scarring patterns are used to identify individuals just as fingerprints are used to identify humans. Blue Whales are identified by their dorsal fin and the distinct mottled coloration on their sides and back. Knowing the individual whales means we can follow their movements and record their behaviors.

New diving technology allows divers to breathe underwater without releasing bubbles. Because many whales release bubbles as a threat or when they are alarmed, a diver who releases bubbles likely disturbs the whale and therefore alters its behavior. Since researchers have been using these kind of diving units, called rebreathers, they have been able to observe male Humpbacks as they "sing." With standard scuba gear, the bubbles alerted the whale and it would immediately stop singing. The challenges of studying whales in the wild are great, but, as we develop and apply new technology, we are able to learn more about the whales and their environment.

Male and female Killer Whale dorsal fins

INTRODUCTION

Whale Lore: from Legends to Modern Mythology

For many cultures, whales have been powerful symbols of creation, perfection and even prophecy. Perhaps the earliest known depiction of a cetacean is found in Norway, where an ancient carving of a Killer Whale lingers on the coastal rocks; it is believed to be about 9000 years old. The ancient Greeks recognized the intelligence and friendliness of dolphins, and they regularly used dolphin motifs in their sculptures, frescoes, pottery, coins and paintings. Certain accounts of Greek mythology attach great meaning and spirituality to the dolphin. In one version of creation, all of life originated within a dolphin; hence the Greek species name *delphys*, which means "womb."

Historically, many indigenous coastal peoples included cetaceans in their mythology in some way. Whales, especially North Atlantic Right, Killer, Bowhead and Humpback whales, are an important part of Native American tradition—they are meshed into the social, religious and imaginative life of the people. In fact, because of the mythological importance of the Killer Whale, it was never actively hunted by Native Americans anywhere in the U.S. The Eskimo and Mi'kmaq also feared the Killer Whale because these curious whales could so easily capsize their small canoes.

The Mi'kmaq and Abenaki people lived in the region of Atlantic Canada and New England long before the arrival of Acadian and British settlers. Their legends tell many stories of Glooscap, a mythical culture hero. One of the many tales about Glooscap explains the great tides in the Bay of Fundy. The legend says that the great Glooscap wanted to take a bath, and so he asked Beaver to build a dam across the mouth of the bay to trap the water. Pleased, Glooscap began his bath. Unfortunately, Whale was angered that the water had stopped flowing to her domain, and she came to the dam and poked her head over it to question Glooscap. Glooscap did not wish to anger Whale, so he got up and went back to land and instructed Beaver to take down the dam. Whale could not wait for Beaver to take the dam down, so with a stroke of her

On the West Coast, indigenous peoples so revered Killer Whales that they are commonly depicted on totem poles.

31

INTRODUCTION

mighty tail she broke the dam and the water came rushing out. To this day the water still rushes in and out of the Bay of Fundy.

Although Native Americans respected whales, many species were prized for their meat and oil. Indigenous peoples believed that whales, dolphins and porpoises lived in underwater villages similar to those of humans. On occasion, the chief whale would order an individual to allow itself to be caught by humans, to give food to the people in need. In gratitude, these people would throw the bones and intestines back into the water, in the belief that these remains would sink back to the whale's village and be renewed into another life.

In recent centuries, many humans feared whales, and contact with them was limited either to hunting or to occasional events of strandings. Mariners told outlandish stories of attacks by great sea creatures that spouted a toxic substance capable of peeling away your skin and blinding your eyes. Artists engraved terrible scenes of freakish giants capsizing boats and lashing out with their tails. Such perceptions were later replaced by phrases like "gentle giants" and stories of whales and dolphins saving human lives. These stories may be true, but the value and appreciation of cetaceans is not dependent solely on their interaction with humans. Indeed, the truth about whales lies somewhere in between the two extremes: they are not quite the gentle and obsequious giants that many people believe them to be, but they are certainly not evil behemoths that lurk beneath our boats, ready to strike.

Whale mythology is not limited to the historical perceptions and accounts of these great creatures. Throughout modern literature and art, whales have been exemplified, celebrated and somewhat idealized. Some famous works include Herman Melville's *Moby Dick*, Hollywood's *Star Trek: The Voyage Home* and *Free Willy*, television's *Flipper* and even Disney's *Pinocchio*. Such stories and films continually add to the rich tradition of whale lore.

Today, the perception and treatment of whales varies considerably among the different people of the world. Many people view whales as a superb opportunity to study mammalian adaptation, whereas others see whales as a source of entertainment. Many societies hunt whales for meat and profit; others consider whales to be creatures of inherent value outside of human valuation schemes.

Regardless of how we look at whales, we share an increasingly intricate relationship with them. Whales, dolphins and porpoises now attract millions of people each year to coastal areas throughout the world. Their special intrigue is a result of their intelligence and sheer magnificence. Watching a Humpback Whale breach and fall back into the water with a tremendous splash or seeing a dolphin's eye turn to look at you as it bow-rides along your boat are extraordinary encounters that leave behind a feeling of contact with very special creatures. Cetaceans demand our respect as powerful creatures that can be friendly and inquisitive in most circumstances, but rightfully protective and aggressive when they are harassed.

INTRODUCTION

Whales in Peril

Understanding Whaling

For centuries, whaling was considered a heroic profession that provided many necessary raw materials for a growing society. Whale parts were made into a staggering array of goods, making a list so long and composing so many unexpected products that it is no wonder the whaling industry was lucrative.

Whale oil was used in lamps, candles, lipsticks, lotions, soaps, shampoos, cooking goods, ice cream, crayons, glycerin (for explosives), machinery lubricants, leather processing, varnishes, adhesives—the list includes virtually everything that requires oil in some way. Baleen from whales has been used for only slightly fewer things, including, but not limited to, umbrella ribs, corsets, dress hoops, whips, window shutters, fishing rods, hairbrushes and shoehorns. Whale meat and blubber was—and still is—eaten by many people; teeth were used as ivory for artworks or piano keys; bones were used for building structures or were ground down to make fertilizer; tendons were used for string and in place of catgut; whale skin was tanned to make leather, and the intestinal linings were stretched and used as transparent window panes. Even ambergris, a hard and waxy digestive by-product that forms in the gut of a Sperm Whale, was used as a fixative for perfumes and was once valued at over $800 per pound.

Most people who hunted whales made a tremendous effort to use every part of the whale and leave nothing to waste. The major difference in usage between groups of people was that some cultures were subsistence whalers and required whales for their survival, whereas other cultures were commercial whalers and used whales

Explosive harpoon

INTRODUCTION

for profit. Even now, people in certain countries still hunt whales for sustenance, and others still hunt whales for profit.

The evidence of a successful whaling industry is indisputable. Over the last two centuries, millions of whales have been killed—so many, in fact, that several species may never recover. In the 1940s, when whale populations were at their lowest, no one could deny that if whaling was not dramatically curtailed, there would be no more large whales. From sheer necessity, the International Whaling Commission (IWC) formed in 1948 and began the first management strategy for the industry and the countries involved in whaling. Unfortunately, the quotas and agreements developed by the IWC were never adhered to, and whaling continued. Moreover, even if countries that reported their annual whale takes to the IWC, most figures were as low as five percent of the actual number of whales that they killed.

A strong argument can be made that the small reduction in whaling that occurred up until the early 1990s happened not because of a global effort to save the whales, but because of a combination of two circumstances: primarily, that the whale populations were decimated, and catches of hundreds of whales each month were no longer possible; secondarily, that owing to the evolution of society, the demand for whale products decreased. Electricity greatly reduced the need for whale oil as a lighting agent, and plastic replaced baleen in many of its uses (and corsets, thankfully, went out of fashion).

In view of the large number of uses for whale products, and though it may not be condonable, it is at the very least understandable that the industry heightened to the point of over-exploiting the resource. After 1925, the combined effect of explosive harpoons, high-speed whaling vessels and floating whale-processing factories marked the peak of commercial whaling. The catch sizes over the next few decades alone were staggering, especially in Antarctic waters. The Fin Whale, previously too fast for the whaling vessels to catch, took the brunt of the onslaught, and nearly 750,000 were killed. Other large whales were taken as well: more than 360,000 Blue Whales, 200,000 Humpback Whales, 200,000 Sei Whales, nearly 400,000 Sperm Whales and more than 100,000 Minke Whales. These numbers represent just a short period in whaling history when whaling vessels could finally catch the faster whales. In the 1800s, the numbers were equally high for the slower whales, such as the right whales and the Bowhead Whale.

Only in the past 20 years has there been a real and obvious reduction in the number of whales killed annually, and the reduction is attributable to the efforts of conservation groups and a worldwide awareness of the plight of whales. Unfortunately, many species are far from safe, and their numbers are far from stable. The North Pacific Right Whale is undeniably the most endangered large whale—only a few hundred survive, and most biologists fear it is too late for this species.

INTRODUCTION

Whaling Along the East Coast

The history of whaling is long and diverse. The earliest whales hunted were Gray Whales and Humpback Whales. Putting a date on the first whaling event is nearly impossible because people were known to eat small whales as early as 6000 years ago. Whether these whales were found stranded or were actually hunted is unknown. Almost certainly, indigenous people hunted whales as early as 2000 years ago.

Prior to the 18th century, American colonists would hunt right whales near shore using small boats. The Gray Whale once inhabited the waters of the east coast, but it disappeared from these waters by the early 1700s. Although the cause of its extirpation is not known for sure, it is likely that whaling is at least partly responsible. Gray Whales inhabit shallow coastal waters, making them easy targets for the early whalers. By the mid-1700s, whaling in New England was a big business, and whaling vessels would depart from Nantucket on voyages lasting months or years. Sperm Whales were targeted because the spermaceti oil made smokeless and odorless candles. By the early 1800s, New England was considered the whaling center of the world, and New Bedford, in Massachusetts, was the fast-growing capital. After 1850, whaling started to decline as kerosene, a superior fuel, became available. The second blow to whaling came from the development of plastics, which crippled the market for baleen.

Today, a few countries are still actively whaling, and some do so illegally. Fortunately, the numbers of whales taken annually are nowhere near historic records, but many cetacean species cannot tolerate more losses. Additionally, some Native American groups are allowed to hunt limited numbers of whales. The Makah of Washington State's Olympic Peninsula, for example, were recently permitted to kill up to five Gray Whales per year. Farther north, some Native Americans in Alaska have been granted similar rights or are in the process of receiving this permission.

Gray Whale

INTRODUCTION

Environmental Concerns

To simply criticize whaling as ecologically unsound is to impose a modern awareness on an industry whose products were historically in great demand. The industry supported and initiated many advancements in society and technology, and even the expansion of the Industrial Revolution was facilitated by whale-oil lubricants. Ironically, the whaling industry of North America made extensive contributions to whale science, and it is this core of knowledge about whales that now helps us protect and understand them. We have made many efforts to save the great whales from further decline as a result of whaling, but other, more subtle, threats exist that may be as dangerous to whales as the harpoon.

A sad fact is that the oceans, which cover 70 percent of the planet, are being used as dumping grounds for a plethora of wastes and by-products produced by modern civilization. Far too many countries have intentionally and accidentally dumped harmful wastes into the oceans, believing that the water will somehow hide or deal with the garbage that no one else will accept. But the oceans are not capable of sanitizing or hiding the billions of tons of toxic waste, agricultural chemicals, crude oil and plastic debris that are carelessly dumped into them, and these harmful substances are now affecting water quality and marine wildlife health around the world.

The magnitude of the problem is difficult to grasp. Many countries dump billions of tons of untreated wastewater into the oceans. There are approximately 65,000 chemicals that are commonly used in industry worldwide, and all of them end up in the oceans in some form or another. Chemical wastes—including DDT, organochlorides and heavy metals—stay in the environment for a long time. High concentrations can cause the death of sea animals immediately, whereas low concentrations can accumulate in their tissues and cause illness, disease and, ultimately, death. Whales that eat small fish or plankton with toxic or chemical waste in their bodies end up storing these harmful substances in their blubber. In the worst cases, dolphins that have died from chemical poisoning and washed up on shore are so laden with pollutants that they themselves are classified as toxic waste. Also, a female's milk, which is made from the body's fat reserves, can be so full of organochlorides and other contaminants that the nursing calf dies. This kind of poisoning of the ocean and marine creatures is extremely insidious and difficult to remedy. It is essential to the health of the oceans that people are educated about how harmful wastes can be, and about how to prevent the discharge from occurring.

One kind of pollution for which every person is responsible is plastic bags. Discarded plastic bags float in the water column and so closely resemble jellyfish and squid that many whales gulp them down. Eventually, a whale's digestive system becomes so blocked by plastic bags that it

INTRODUCTION

dies of starvation. Originally, when creatures that eat jellyfish or squid—such as sea turtles and whales—were found dead on shore, the cause of their deaths was unknown. It was only after autopsies that plastic bags were found to be the problem. Millions, or maybe even billions, of plastic bags are discarded every day, many of which will end up floating in the oceans.

There is no easy answer to the problem of marine pollution. Of course, the careless discarding of harmful products into the water must be curtailed, but the existing pollution in the water is not so easily cleaned up.

Fishing nets are also responsible for the deaths of innumerable cetaceans each year. The gill nets commonly used for fishing often catch small whales, dolphins and porpoises. Once caught in a net underwater, a cetacean drowns because it must breathe air. When the nets are pulled in, the bodies of the drowned cetaceans must be untangled from the net. Eventually, a net is so full of holes that it is irreparable, and it is often discarded into the water. A discarded "ghost net" continues to catch sea creatures and small cetaceans while it floats freely through the ocean, until it sinks to the bottom because of the weight of the bodies trapped in its mesh. Large whales may become entangled in ghost nets too, and they end up towing the net with them wherever they go. The net eventually digs into the whale's flesh, causing life-threatening injuries and preventing it from feeding properly.

If you are out whale watching and you see a whale entangled in a fishing net, the best course of action is to report this whale and its location to an emergency hotline. The telephone numbers of people who will help a whale in trouble are listed on pages 159 to 162 of this book.

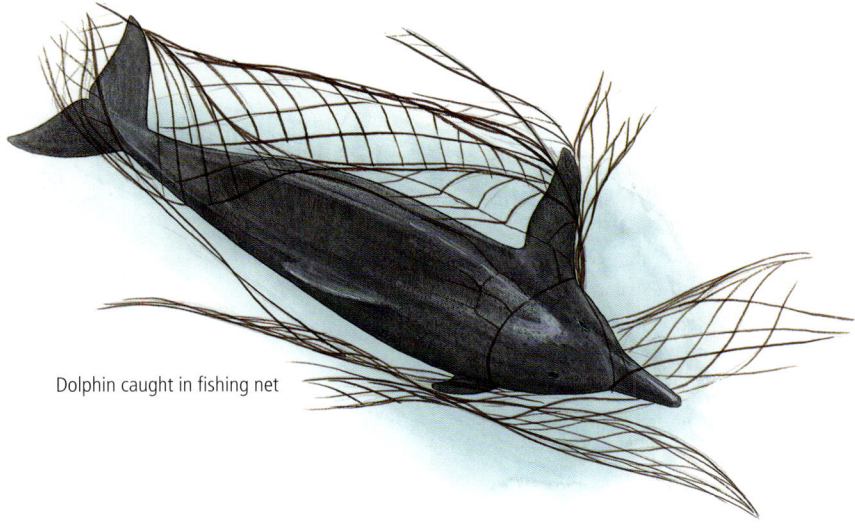
Dolphin caught in fishing net

INTRODUCTION

Whale Watching

Whale watching has become a worldwide activity that attracts millions of people each year. The east coast has many famous sites where whale watchers can easily see North Atlantic Right, Humpback, Fin, Bryde's and Northern Minke whales. Other species that can be seen unexpectedly at any time include Killer, Sei and Sperm whales. Several dolphin species are regularly seen in certain locations. The preceding species are the most frequently seen cetaceans, but remember that every species in this book has been found within the waters from Maine to Florida. Keep your eyes open for the rarer whales—positive sightings are worth reporting to local research stations.

The east coast of the United States has some of the most exciting places in the world for viewing and studying whales. Despite past whaling and current heavy boat traffic, the natural harbors and food-rich waters of our coast form prime habitat for many species of whales. Several species of whales, most notably the North Atlantic Right Whale, perform their long migration along this coast, and some of the best-known and most-studied populations of Fin and Humpback whales migrate through these waters. In addition to migrating whales, many species, especially toothed whales, make these waters their home year round.

Although it is true that you may encounter whales throughout the east coast, some places are better than others to find whales. The map on the facing page illustrates the best sites for whale and marine mammal watching, which usually correlate to good availability of tours. There are even places on land where you can watch whales as they migrate past. The Great Dune at Cape Henlopen State Park in Delaware offers a perfect vantage point for shore-based whale watching. With respect, the places where whales are sensitive to disturbance or places that are ecologically fragile have not been included. Nevertheless, some of the best whale watching sites in the world are here, ready for you to explore.

North Atlantic Right Whale

INTRODUCTION

Top Whale Watching Sites

1. Bar Harbor
2. Boothbay Harbor
3. Portsmouth
4. Boston
5. Cape Cod
6. Nantucket
7. Newport
8. Montauk
9. Cape May
10. Delmarva Peninsula
11. Virginia Beach
12. Cape Hatteras to Cape Lookout
13. Murrells Inlet
14. Gray's Reef National Marine Sanctuary
15. Brunswick
16. Blue Spring
17. Fort Pierce
18. Miami

INTRODUCTION

Sea floor topography plays an important role in whale abundance and diversity. Although we visualize the coast as the edge of the land, the continental shelf actually continues as much as 250 miles out to sea. The Atlantic coast is relatively low, compared to the hilly and mountainous Pacific coast. Low coastal areas of the mid and southern Atlantic coast, Gulf coast and the continental shelf that extends into the water are called the Atlantic Plain. North of Cape Cod is geologically distinct, and this region, including the continental shelf there, is called the Appalachian Highlands. From Maine to Florida, however, the continental shelf is a vast region of relatively shallow water. Most species of dolphins and some large whales such as the Humpback Whale and the North Atlantic Right Whale prefer these shallow shelf waters. Beyond the edge of the continental shelf is the deep ocean, where species such as the beaked whales, sperm whales and some of the large rorquals are encountered. Ideally, the best whale watching areas are those that have access to both shelf water and off-shelf deep ocean. Whale watching trips from Cape Cod, for example, can access waters beyond the continental shelf. The Hydrographer Canyon is at the shelf edge near Cape Cod, and it can be reached during an all-day boat trip. Possible species to see in waters above the canyon include the Sperm Whale and various beaked whales.

Another factor that strongly influences marine life is the Gulf Stream. The Gulf Stream is a warm ocean current that originates off the tip of Florida and moves northward along the coast toward Canada. "Stream" is a misnomer, however, as the current is extremely powerful, and its volume is far greater than all the rivers emptying into the Atlantic. It is one of the strongest ocean currents in the world, second only to the Antarctic Circumpolar Current. Sperm Whales are strongly associated with the Gulf Stream, as are beaked whales and certain dolphins.

There are many choices to make if you are to have a successful whale watching adventure. The time of year is important: catching the peak of migration makes for the best chances of seeing certain species. Humpback, Fin and North Atlantic Right whales are famous seasonal visitors in this region. About 1000 Humpback Whales spend the summer in the Gulf of Maine. Other species, such as dolphins and Killer, Minke, Sperm and Long-finned Pilot whales, are uncommon but can be seen year-round. If you want to take a boat tour, you can choose between a day trip or a multiple-day cruise. Costs vary depending on how long of a trip you are taking.

Always take the time to learn about the person operating the tour. A tour leader should be knowledgeable about whale biology and be up to date on current locations of different whales. It is essential that the tour operator is responsible and behaves ethically around whales. Like people, whales have personalities and personal space requirements, and boaters who harass them (even unknowingly) can

INTRODUCTION

be extremely disruptive to the animal and can cause unnecessary stress and harm. In the worst cases, a whale that is badly harassed may show aggression toward boaters and can suffer stress or injuries. In the best cases, a whale may be comfortable with your presence and approach your boat on its own accord to get a better look at you. If you are the skipper of your own boat, know the basic guidelines for approaching whales. Although the following guidelines may appear to be strict, they ensure the safety of the whales and the boaters:

- minimum approach distance is 300 feet
- approach and leave whales at "no-wake" speed from the side or behind (never from ahead), with no sudden changes in speed or direction
- if a whale approaches, put the engine in neutral and allow it to pass
- never follow whales or herd, drive or separate them, particularly mothers and calves
- have only one vessel at a time at the minimum approach distance
- maintain a time limit of 30 minutes per vessel at the minimum distance
- swimmers should not approach within 150 feet.

The Marine Mammal Protection Act of 1972 prohibits the harassing, capturing or killing of any marine mammal. Anyone caught breaking this law is subject to fines of up to $25,000 and/or imprisonment. Most tour operators are knowledgeable and responsible and show genuine concern for the welfare of whales. If you encounter people who are not following proper conduct around whales, please speak up. Let them know they are interfering with and possibly harming the whales. If everyone acts responsibly, the whales will probably stay in the vicinity for everyone to enjoy.

The industry of whale watching is more than just the intrinsic pleasure of seeing these great creatures. The top whale watching regions along the east coast bring in millions of dollars every year, and, economically speaking, whale watching is worth far more than whaling. To look at whale watching in these monetary terms is important if whale conservation is to be politically and economically acceptable. However, attaching dollar figures to whale watching is superficial at best. How can we put a dollar figure on awe-inspiring experiences in the wild? As every whale watcher knows, the true value is in the extraordinary and breathtaking encounter with a wild whale.

Fin Whale

41

INTRODUCTION

About this Book

No matter where you are learning about whales, whether aboard a whale watching boat on the ocean or in the comfort of your own home, keep this book handy for quick reference about whales and other marine mammals. All the species you might encounter off the east coast are illustrated and described within this book. As well, historical and biological information is presented to give you a better understanding about whales in general.

About 34 whale species occur in the waters off the east coast. Of these, seven are baleen whales (mysticetes) and 27 are toothed whales (odontocetes). The most famous and commonly seen cetaceans in this region are the North Atlantic Right Whale, Humpback Whale, Fin Whale, Northern Minke Whale, Long-finned Pilot Whale, Bottlenose Dolphin and Harbor Porpoise, but less common cetaceans may be encountered at any time.

The whales of the east coast are presented here in a generally accepted sequence that places the more closely related whales near one another. The scientific names and the ordering of the groups (genera) and species follow the "Revised checklist of North American mammals north of Mexico, 2003" (Occasional Paper No. 229, Museum of Texas Tech University, by Jones, Hoffmann, Rice et al. 2003). The section describing the other marine mammals in the region is similarly ordered, but is placed after the section on whales. The common names used in this book are widely accepted and frequently used names. Owing to discrepancies in the common names of many species, however, some species may be found under different names in other texts. Wherever applicable, the alternate common names are included in each entry as well as in the index.

The quick reference guide on pages 4 to 6 illustrates all the species in the book. It may help you identify species and groups at a glance, and it leads you to the detailed description of each species. The blows and dive sequences for selected species are also illustrated together on pages 23 and 24. The glossary (pages 156 to 158) defines some of the specialized terms that are used in talking about whales, and it also includes labeled diagrams of the main cetacean bodyparts.

Each species account includes a "Status" entry that gives information on the current conservation status of the species. Descriptions such as "endangered," "threatened" and "vulnerable" are official designations made by the U.S. Fish and Wildlife Service (USFWS) and are defined in the glossary. Other descriptions appearing in the book, such as "common," "stable" or "declining," are used in accordance with current local research.

Each species account also includes a range map. Dark blue water indicates the

INTRODUCTION

known range for that species. You may find the range maps helpful when trying to determine the identity of a whale in the wild.

It should be noted that the lifecycles of many whales are described in terms of the animal's size instead of its age. Because of our limited understanding of cetaceans, we are unable to estimate an individual's age for most species, whereas length can be readily measured in the field.

Because some whales are so very large, it can be hard to get a good visual idea of their size. To help you with this task, below the size measurements we show a scaled comparison between an average-sized individual of each species and either a human diver (about 6 feet tall) or a school bus (about 35 feet long).

Sample range map

These silhouettes will help give an idea of scale when considering the size of a whale.

The "Similar Species" section in each species account lists other whales or marine mammals that could easily be confused with the animal you are reading about. Be sure to check these other species before you decide on the identity of an animal in the wild.

Once you have learned to identify whales and you are familiar with their natural history, enjoy them! Whales are marvelous creatures, and watching them can be a thrilling and memorable experience.

Killer Whale

43

INTRODUCTION

Whale Tales & Record Breakers

This section attempts to clarify some of the frequently asked questions about whales and whale extremes. Keep in mind, however, that these answers are not absolute. As we learn and discover more about the marine world, we have to continually update our record sheets.

How many different whales are there in the world?
There are currently 88 known species of whales, dolphins and porpoises.

What is the largest whale in the world?
Blue Whale: maximum record 110 ft.

What is the smallest whale in the world?
Franciscana (river dolphin): 4½–6 ft.

What is the heaviest whale in the world?
Blue Whale: maximum record 200 tons.

What is a Whale Shark?
A Whale Shark is a shark, not a whale. The modifier "whale" is used because this shark is very big (up to 60 ft long).

What is the fastest whale in the world?
Atlantic White-sided Dolphin: 32 knots.

What is the longest-lived whale?
Bowhead Whale: more than 200 years.

What is the deepest-diving whale?
Sperm Whale: deepest record 10,000 ft.

INTRODUCTION

How long can a whale hold its breath?
Sperm Whale: confirmed record 2 hours, 18 minutes.

What is the longest migration made by a whale?
Gray Whale: round trip (one year) 12,400 mi.

How much food can a whale eat in one day?
Blue Whale: up to 7 tons.

What is the longest pregnancy for a whale?
Killer Whale: 15 to 16 months.

What is the rarest whale in the world?
Yangtze River Dolphin (China): fewer than 100 remaining.

What is the loudest sound made by a whale?
Blue Whale: 188 decibels, measurable for at least 1500 mi underwater.

Which whale has the most baleen plates?
Fin Whale: 360 plates (on average).

Which whale has the longest baleen?
Bowhead Whale: 14 ft. One disputed report of 19 ft.

Which whale has the most teeth?
Spinner Dolphin: 224.

Which whale has the longest teeth?
Narwhal: a single tusk can reach 10 ft long. One in 500 males has two tusks.

Baleen Whales

Rorqual Family (Balaenopteridae)

The rorquals number only six species worldwide, and all of them may be seen off the east coast. They include some of the largest whales on earth—the Blue Whale is reputed to be one of the largest animals that has ever lived. The name "rorqual" is derived from the Norwegian word *rorhval*, meaning "furrow," which refers to the pleats, or folds, in the skin of the throat. These throat pleats expand during feeding, allowing the throat to distend to an enormous, balloon-like shape when the whale gulps a massive volume of food-rich water into its mouth. The whale does not swallow the water; instead, the pleats contract and the tongue moves forward to push the water out through the baleen. Any crustaceans or fish are trapped inside the mouth and swallowed whole. The only whales outside the rorqual family that have throat pleats are the Gray Whale and the beaked whales, but their pleats are relatively few and have greatly reduced efficacy.

Rorquals are easily identified by their pointed snouts, flattened heads, long and slender bodies and relatively small dorsal fins, which sit about two-thirds of the way down the back. When a rorqual's mouth is open, the short baleen, which is continuous around the front of the jaw, is visible.

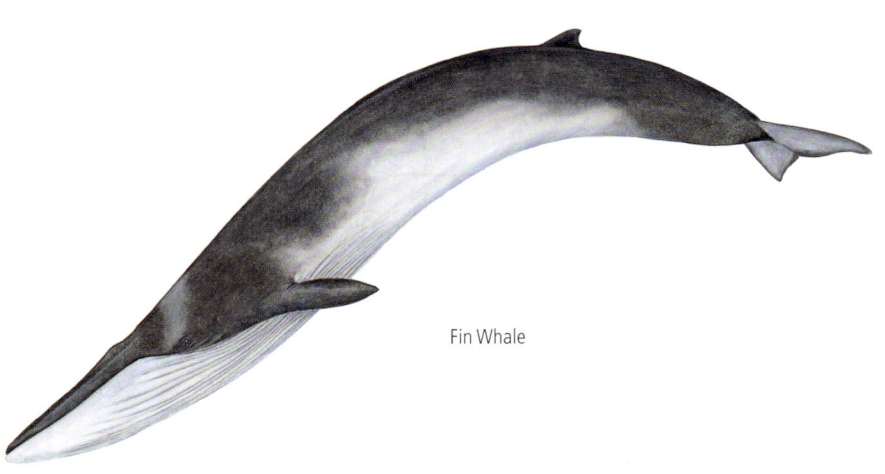

Fin Whale

Right Whale Family (Balaenidae)

This family includes two kinds of whales, the right whales and the Bowhead Whale. The Bowhead Whale is found only in arctic and subarctic waters. Right whales have a much larger distribution.

The endangered right whales were once found throughout the world's cold and temperate oceans. Their total population now is probably only a few thousand individuals. There is much argument about whether there are one, two or three species of right whales. Many scientists believe there is only one species worldwide, *Eubalaena glacialis*, while others feel strongly that this name represents only the North Atlantic population, and that the Southern Hemisphere has a different species (*E. australis*). The third species (*E. japonica*) inhabits the north Pacific from the Sea of Okhotsk to the Bering Sea.

Among the most distinct features of a right whale are the growths over its face and head. These callosities are masses of keratin topped with barnacles and teeming with whale lice. The whale lice may be pink, white, orange or yellow, and they give the callosities their unique color. The origin of these callosities is intriguing—they grow in all the places on the whale's head where a human would have hair. Specifically, they are found on the "chin," the upper "lip" and above the eye. Keratin makes up both fingernails and hair in other mammals, and this fact may explain the origin of callosities in right whales.

North Atlantic Right Whale

Northern Minke Whale
Balaenoptera acutorostrata

The Northern Minke Whale, the smallest of the rorquals, is relatively common along the east coast. Despite their numbers, minke whales are elusive and difficult to spot because they spend less time at the surface than other whales—they usually take only five to seven breaths before they go under again. As well, minke whales rarely make a visible blow, and they usually travel singly—one little dorsal fin against an ocean of waves is easy to overlook. Yet there are cases where a minke whale has surfaced right next to a boat, much to the surprise of the people aboard. Such occurrences are fleeting; the whale rapidly takes a breath and disappears as quickly as it came.

Several researchers who have studied minke whales consider them extremely intelligent and adaptable. Minke whales are known to look for flocks of gulls, murres or auklets on the water—a flock of feeding birds advertises the whereabouts of a school of fish, making the fish an easy lunch for the whale. Although minke whales in equatorial and southern waters frequently use this tricky technique, the behavior is not as common in northern areas.

The study of minke whales is still in its early stages, owing partly to their elusiveness, but also to their wide distribution. People wanting to study right whales, for example, know exactly where to find them, but minke whales can be anywhere, and we have yet to determine their behavior patterns. It is known that many minke whales migrate between warm and cold waters each year, but some are believed

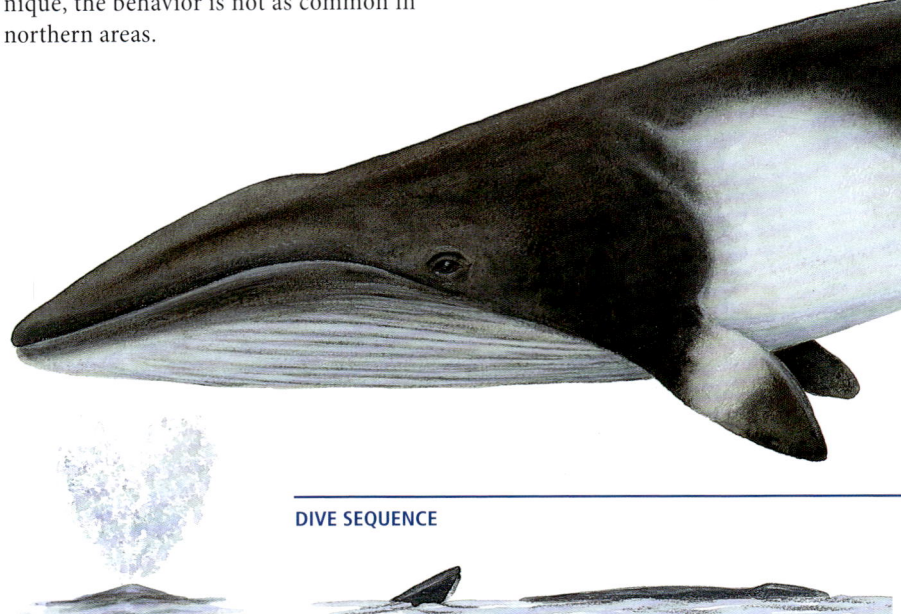

DIVE SEQUENCE

RORQUALS

RANGE: widely distributed throughout the world: in summer it is common in cold polar waters; in winter, migrates to more temperate waters.

OTHER NAMES: Common Minke Whale, Piked Whale, Sharp-headed Finner, Little Finner, Lesser Finback, Lesser Rorqual.

STATUS: insufficient data worldwide; locally common.

ADULT LENGTH: up to 35 ft; average 27 ft.

ADULT WEIGHT: up to 15 tons; average 10 tons.

BIRTH LENGTH: 8–9 ft.

BIRTH WEIGHT: 770 lb.

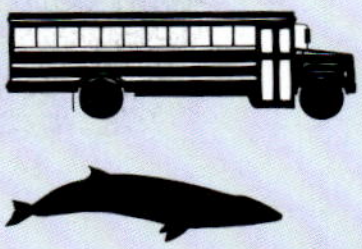

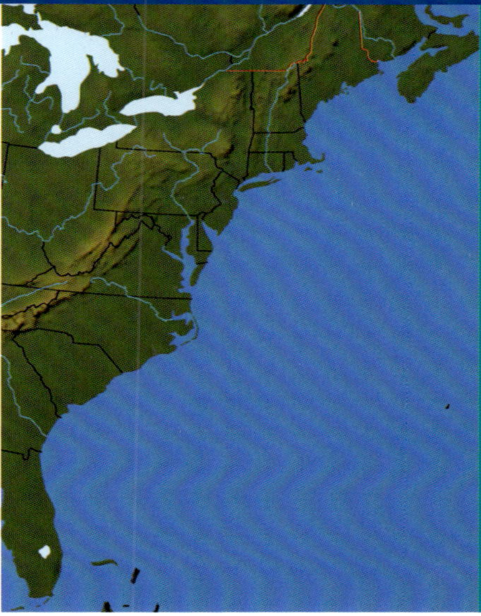

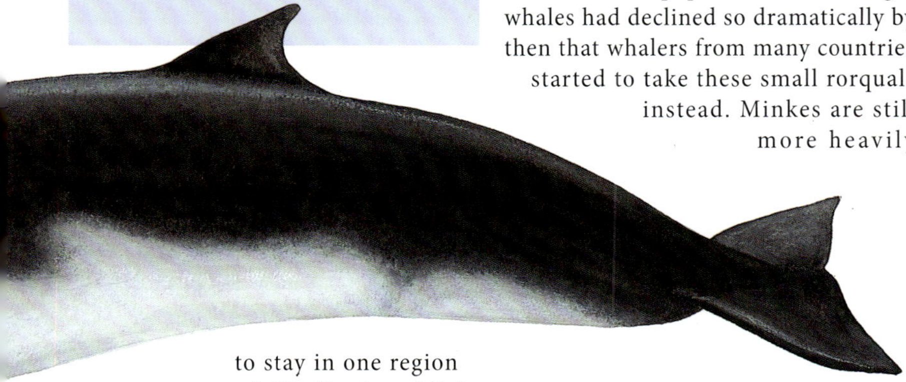

BALEEN WHALES

In the 1980s, minke whales were the most heavily hunted baleen whales in the world. The populations of the larger whales had declined so dramatically by then that whalers from many countries started to take these small rorquals instead. Minkes are still more heavily hunted than almost any other baleen whale, but new regulations and international efforts have protected them in many parts of their range.

to stay in one region year-round. The Northern Minke Whale's winter mating and calving waters in the North Atlantic are not well studied. New advances in satellite and radio taps, however, may help fill in some of the knowledge gaps.

49

RORQUALS

DESCRIPTION: The Northern Minke Whale is striking, with whorls of white over its dark back and white "epaulets" on its flippers. On some individuals, the entire flipper may be white; on others, the white may be just a narrow band or even nonexistent. This variation in flipper markings is regionally specific, and it might indicate different subspecies. The coloration over the back varies considerably throughout a minke's range: it can be dark slate gray, nearly black, deep bluish or dark brown. A minke whale's undersides are white or nearly white, and there are often swirls of white behind the flippers and again below the dorsal fin. This whale's overall appearance in the water is very sleek and streamlined. Its head is quite pointed, with a prominent splashguard before the blowhole and a distinct single ridge to the snout. Relative to most other rorquals, the Northern Minke Whale has a large, curved dorsal fin. Females are generally larger than males.

Flipper variations

RORQUALS

areas of high food concentration, a hundred or more minke whales may occur together.

FEEDING: Like the larger rorquals, the Northern Minke Whale feeds mainly on krill, but it also eats some other invertebrates and small fish. It commonly lunge-feeds by swimming into a school of krill or small fish and gulping a large volume of food-filled water into its mouth. Its throat stretches like a big balloon, and as the whale closes its mouth and the throat pleats contract, the water is forced out through the baleen but the organisms in the water are trapped in the mouth, and the whale uses its tongue to wipe the creatures off the baleen so it can swallow them.

REPRODUCTION: Minkes probably mate in late winter, but no calving waters have been found for the North Atlantic population, and very little is known about their reproductive cycle. The gestation period is 10 months. The nursing period is short, 3 to 6 months. Healthy females will mate again within a few months of giving birth, resulting in a calf every 12 to 14 months. Male minke whales are sexually mature when they reach about 23 feet in length, and females when they are 24 feet long.

SIMILAR SPECIES: The Sei Whale (p. 52) is almost twice the size and lacks white "epaulets" on its flippers; the Bryde's Whale (p. 54) has 3 ridges along its rostrum.

BLOW: The Northern Minke Whale may make a small, quick, bushy blow that reaches a maximum height of about 9 feet. The blow is rarely visible, especially on warm days. On a very calm day, you may hear the blow rather than see it.

OTHER DISPLAYS: When a minke whale breaches, it comes far enough out of the water to expose its dorsal fin. The whale leaves the water at about a 45 degree angle, and it usually does not twist or turn as it comes down with a tremendous splash. This whale often arches back into the water headfirst, so the breach looks more like a dive than a belly flop.

GROUP SIZE: Minke whales are seen either singly or in groups of 2 to 4 individuals. In

Sei Whale

Sei Whale
Balaenoptera borealis

DESCRIPTION: The Sei Whale resembles the larger Fin Whale (p. 60) and the similarly sized Bryde's Whale (p. 54), and the three species can be hard to distinguish in open ocean. The Sei Whale is mainly dark gray, blackish or bluish gray, with pale gray or whitish undersides, and it may have irregular, pale scars and marks along its body. Its baleen is finer and silkier than that of other rorquals, and its throat pleats are shorter. The dorsal fin is slender, upright and curved on the trailing edge. The flippers are dark on both sides, and they have pointed tips. The tail flukes are small relative to the size of the body. Females are generally larger than males.

BLOW: The Sei Whale makes a blow that is similar to, but smaller than, those of the Blue Whale (p. 56) and the Fin Whale. The blow is generally narrow, but not dense, and it may be up to 10 feet tall. It is sometimes mildly heart-shaped.

OTHER DISPLAYS: Sei Whales breach infrequently and rarely more than once at a time. They leave the water at a low angle and re-enter with a splash.

GROUP SIZE: Sei Whales are seen either singly or in groups of up to 5 individuals. Feeding waters attract larger numbers.

FEEDING: The Sei Whale feeds primarily on small schooling fish or invertebrates such as squid, krill and copepods. Its feeding style is similar to a right whale's: rather than lunge-feeding like the other rorquals, the Sei Whale tends to skim steadily through food-rich water.

REPRODUCTION: Sei Whales mate and bear their young in mid-winter. The

DIVE SEQUENCE

RORQUALS

BALEEN WHALES

RANGE: found in deep, temperate waters worldwide: in summer, may travel to food-rich subpolar waters; for the rest of the year, primarily found at lower latitudes.

OTHER NAMES: Pollack Whale, Coalfish Whale, Sardine Whale, Japan Finner, Boreal Rorqual, Rudolphi's Rorqual.

STATUS: endangered (USFWS).

ADULT LENGTH: up to 69 ft; average 49 ft.

ADULT WEIGHT: up to 32 tons; average 25 tons.

BIRTH LENGTH: 14–16 ft.

BIRTH WEIGHT: about 1600 lb.

Northern Minke Whale (p. 48) is much smaller and usually has white "epaulets" on its flippers; the Fin Whale (p. 60) is larger, has a white lower right jaw and has more and longer throat pleats; the Blue Whale (p. 56) is much larger and has a relatively smaller dorsal fin.

gestation period is 11½ months. Each female bears a single calf. These whales are sexually mature when they are about 95 percent of their total adult length.

SIMILAR SPECIES: The Bryde's Whale (p. 54) has 3 ridges along its rostrum; the

Bryde's Whale

RORQUALS

Bryde's Whale
Balaenoptera edeni

DESCRIPTION: Like many of the other large rorquals, the Bryde's Whale (pronounced broo-des) has a long and slender appearance. This whale can be easily confused with the Sei Whale (p. 52). The best identifying feature is the ridges along the top of the rostrum. Whereas other rorquals have only 1 ridge, the Bryde's Whale has 3 parallel ridges. The overall body color is slate blue or grayish, with some individuals appearing somewhat brown. Along the sides, especially toward the flukes, the skin may appear dappled or scarred. The undersides are much lighter in color, either white, gray or cream-colored. The dorsal fin is small but strongly curved and sits about ⅔ of the way down the dorsal ridge. The flippers are slender and solidly colored, and the flukes are broad, slightly concave and distinctly notched in the middle. The underside of the flukes is usually lighter than the top, sometimes nearly white. Females are larger than males.

BLOW: The blow is tall and thin, and rarely over 14 feet high. The spray disperses quickly and may not be visible at a distance.

OTHER DISPLAYS: When a Bryde's Whale breaches, up to ¾ of its body rises nearly vertically out of the water before it splashes back in. It usually breaches multiple times in a row. This whale may swim so rapidly and with such agility that it has the appearance of a very large dolphin.

GROUP SIZE: Bryde's Whales may be seen singly or in small groups of 3 to 7 members. Good feeding sites can attract up to 30 individuals.

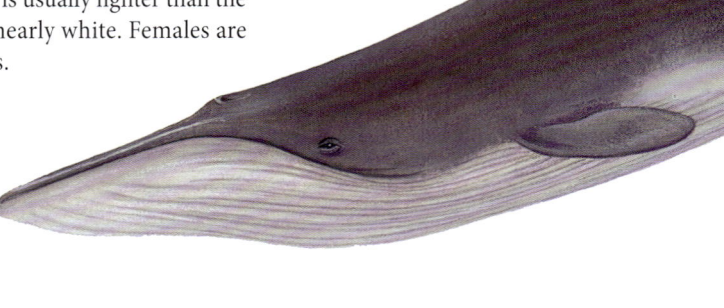

DIVE SEQUENCE

RORQUALS

RANGE: inhabits warm waters (above 68° F) around the world, between 40° N and 40° S.
OTHER NAMES: Tropical Whale.
STATUS: unknown.
ADULT LENGTH: up to 51 ft; average 40 ft.
ADULT WEIGHT: up to 22 tons; average 15 tons.
BIRTH LENGTH: 11 ft.
BIRTH WEIGHT: 1 ton.

BALEEN WHALES

FEEDING: The Bryde's Whale feeds on relatively large shoaling fish as well as copepods such as krill. This whale primarily lunge-feeds.

REPRODUCTION: No distinct breeding season has been defined for the Bryde's Whale, and they may mate at any time of the year. The gestation is about 12 months. A single calf is born, and it suckles for at least 6 months. Sexual maturity of females is reached at about 80 percent of their adult length; males are sexually mature at about 85 percent of their adult length.

SIMILAR SPECIES: All of the other rorquals (pp. 48–65) have only 1 ridge along their rostrum; the Northern Minke Whale (p. 48) is much smaller; the Blue Whale (p. 56) is much larger.

Blue Whale

RORQUALS

Blue Whale
Balaenoptera musculus

Bigger than a dinosaur, bigger than a jet plane, bigger than three school buses—the words we use to describe the size of the Blue Whale hardly do justice to this leviathan. We cannot easily comprehend or describe how large the Blue Whale is, and perhaps this difficulty is reflected in the 18th-century humor used to scientifically name the species. As if in jest, because any word we could use to describe this whale would fall short of capturing its true immensity, the Latin word *musculus*, meaning "little mouse," was used. The Blue Whale is so manifestly not a mouse that we can only smile at the comparison.

The Blue Whale is the most famous record-breaker in the animal kingdom. Not only is it the largest creature alive, and one of the largest there ever has been, but it is also the loudest—its low-frequency pulses have been measured at up to 188 decibels, as loud as a space shuttle engine. At birth, a Blue Whale calf is 23 feet long, which is already larger than at least two-thirds of whales at their full adult length, even though the gestation period—about a year—is pretty typical of baleen whales. The calf grows at an incredible rate: it puts on a staggering 200 pounds a day, nursing 50 gallons of milk from its mother every day, and a weaned juvenile (about eight months old) is about 50 feet long. As adults, the largest individuals may eat up to 4 tons of food a day during summer. An adult's heart is as large as a compact car, and a small child could crawl through the arteries. A male's penis is about 10 feet long.

Another record the Blue Whale holds, unfortunately, is that it is one of the most endangered large

DIVE SEQUENCE

RORQUALS

BALEEN WHALES

RANGE: usually found in open waters; although distribution appears to be worldwide, it is not continuous.

OTHER NAMES: Sulfur-bottom, Sibbald's Rorqual, Great Northern Rorqual.

STATUS: endangered (USFWS).

ADULT LENGTH: up to 110 ft; average 85 ft.

ADULT WEIGHT: up to 200 tons; average 120 tons.

BIRTH LENGTH: about 23 ft.

BIRTH WEIGHT: about 2.5 tons.

whales in the world. At the height of the whaling industry in the late 1800s and early 1900s, Blue Whales were relentlessly pursued. By 1950, the species was declared endangered. In 1966, the International Whaling Commission gave a protected status to the Blue Whale, and it is slowly recovering. The total population is estimated to be about 10,000 individuals, but this number could be as high as 20,000.

Although there is no guarantee of seeing a Blue Whale, there are good chances to see these magnificent creatures in some regions. Although they are rare in the shelf waters of the east coast, they may been seen in summer and fall off Cape Cod. Blues are strongly nomadic, and they do not stay in the same waters for more than 10 days. There is a moderate migration between northern or polar waters in summer and southern or equatorial waters in winter. Although some species of whales fast when they are in their warm winter waters, there is evidence that Blues feed year-round, likely a necessity considering their size.

RORQUALS

DESCRIPTION: The enormous Blue Whale is typically deep slate blue or grayish in color, with variable pale gray or white mottling over the back. Its undersides are lighter, ranging in color from pale blue-gray to white to yellowish. A film of diatoms on the pale undersides results in yellowish bellies on some individuals and gives rise to the alternate name "Sulfur-bottom." The Blue Whale has a broad, flattened head, a large splashguard before the blowhole and a pale throat with 55 to 88 pleats. The relatively tiny dorsal fin is located about ¾ of the way down the body. The flippers are long, slender and darker above than below. The tail stock is very thick. The flukes are narrow and slightly notched, and the tail is very wide from tip to tip—almost ¼ of the body length in width. Females are generally larger than males.

BLOW: The blow of the Blue Whale is distinctive: the narrow column of spray can rise at least 35 feet high, and it can be seen even at a distance of 3–5 miles.

RORQUALS

6 minutes, and then it will dive for 10 to 15 minutes.

GROUP SIZE: Blue Whales are typically seen singly or in pairs. Even when food-rich water attracts high densities, they remain as singles or pairs.

FEEDING: The Blue Whale specializes in eating krill—shrimp-like creatures that are less than ½ inch in length—and it rarely eats anything else. A Blue Whale usually rises up into a school of krill from below, gulping in hundreds of gallons of seawater. The water is squeezed out through the baleen either rapidly or over the course of several minutes.

REPRODUCTION: Gestation lasts 11 to 12 months, and the peak calving period appears to be during winter, when the whales are at low latitudes. Calves are weaned when they are about 8 months old. Both sexes reach sexual maturity when they are at least 80 percent of their adult length—about 10 years old.

SIMILAR SPECIES: The Fin Whale (p. 60) is smaller and has a larger dorsal fin; the Sei Whale (p. 52) is much smaller and has a larger dorsal fin; the Bryde's Whale (p. 54) is much smaller and has 3 ridges along its rostrum.

OTHER DISPLAYS: Not surprisingly, the Blue Whale is not acrobatic, but a juvenile may breach and make a tremendous splash as it falls back into the water, on either its side or its belly. An adult's dive sequence is marked initially by the tall blow; then the whale's long back rolls over the surface. The dorsal fin appears well after the head has disappeared and the blow has dispersed. During its dive sequence, a Blue Whale may surface several times over a period of up to

Fin Whale

Sei Whale

Fin Whale
Balaenoptera physalus

DESCRIPTION: The unusual Fin Whale has a striking asymmetry of coloration on its head: the right side of the outer lower "lip" and the baleen on the right side are whitish, whereas the counterparts on the left side are gray. Oddly enough, the mouth cavity and tongue are gray on the right and white on the left. There is a distinct pale gray, V-shaped mark behind the head. The overall body color is dark gray, silvery gray or brownish black, and the undersides are white. The dorsal fin is small and slanted backward. The flippers and flukes are dark on top and white below. There is a distinct ridge on the back from the dorsal fin to the flukes. Females are generally larger than males.

BLOW: The Fin Whale's narrow blow is quite dense and is 13–20 feet high, making it visible from far away.

OTHER DISPLAYS: Fin Whales may make spectacular breaches, leaving the water at 45 degrees, but they rarely show their dorsal fin. They may twist in the air as they come down with a thunderous splash.

GROUP SIZE: Fin Whales are occasionally found singly or in pairs, but more typically they are found in groups of 3 to 10 individuals.

FEEDING: The main diet of the Fin Whale includes schooling fish, krill, squid, copepods and other invertebrates. Lunge-feeding, during which the whale often turns on its right side, is commonly seen.

REPRODUCTION: Fin Whales mate in late winter. About 11 months later, 1 calf is born. Males are sexually mature at 90 percent of their adult length; females at 75 to 80 percent.

DIVE SEQUENCE

RORQUALS

BALEEN WHALES

RANGE: found worldwide in temperate and subpolar waters.

OTHER NAMES: Common Rorqual, Finback, Razorback, Herring Whale, Finner.

STATUS: endangered (USFWS).

ADULT LENGTH: up to 89 ft; average 70 ft.

ADULT WEIGHT: up to 140 tons; average 80 tons.

BIRTH LENGTH: 20–22 ft.

BIRTH WEIGHT: about 2 tons.

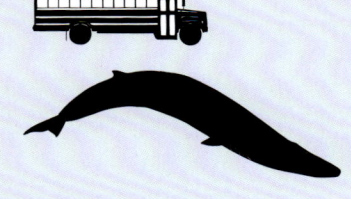

SIMILAR SPECIES: The Blue Whale (p. 56) is larger and has a smaller dorsal fin; the Sei Whale (p. 52) is smaller, has fewer and shorter throat pleats and does not have a white lower right jaw; the Bryde's Whale (p. 54) is smaller and has 3 ridges along its rostrum.

Blue Whale

Sei Whale

Humpback Whale

Megaptera novaeangliae

Humpback Whales are renowned for both their extensive migrations and their haunting songs. They are commonly seen by whale watchers, and they seem to enjoy performing for their boat-bound admirers. Other than brief bouts of fighting between males during the winter breeding season, Humpback Whales have gentle and docile natures.

In summer, Humpbacks feed and socialize off New England. In winter, they migrate south to Caribbean waters, where they mate and calve. There are six major feeding aggregations of Humpback Whales in the North Atlantic. One of these is found in the Gulf of Maine, and Stellwagen Bank just north of Cape Cod is one of the best places in the world to view feeding Humpback Whales. By September, the whales begin their migration south. In general, Humpbacks migrate as much as 5000 miles between high latitude summer feeding waters and low latitude mating and calving waters. Although feeding and mating waters are distinct between different populations of Humpbacks, there are individuals that switch from one population to another.

DIVE SEQUENCE

RORQUALS

RANGE: found worldwide, from polar to tropical waters; migrates seasonally, so some regions have higher densities than others.
STATUS: endangered (USFWS); locally common.
ADULT LENGTH: up to 62 ft; average 45 ft.
ADULT WEIGHT: up to 53 tons; average 30 tons.
BIRTH LENGTH: 13–16 ft.
BIRTH WEIGHT: 1–2 tons.

BALEEN WHALES

The song of the Humpback

Whale is one of the most impressive sounds in the animal kingdom. A Humpback can sing for a few minutes to half an hour, and an entire performance can go on for several days, with only short breaks between each song. The songs are composed of trills, whines, snores, wheezes and sighs, and when sung in a repeating pattern, the result is mysterious and hauntingly beautiful. Although the true meaning of the Humpback's song eludes us, we do know that only the males sing and that they perform mainly during the breeding season. Many animals use sound to attract the opposite sex, which may be the primary function of the Humpback's songs.

Male Humpbacks become very aggressive toward one another and battle to determine dominance in their winter breeding waters. A dominant male becomes the escort to a female with a calf. Presumably, a female with a calf is one that is, or will soon be, receptive to mating.

RORQUALS

DESCRIPTION: The Humpback Whale is slightly more robust in the body than are other rorquals. The overall body color is either dark gray or dark slate blue, and the undersides may be the same color as the back or nearly white. The slender head bears many knobs and projections around the snout, and the mouth line arches downward to the eye. There are 12 to 36 pleats on the pale throat. The flippers are long and knobby and have a varying pattern of white markings. The tail flukes are strongly swept back and have irregular trailing edges. Like the flippers, the flukes have unique white markings that can be used to identify individual whales. The dorsal fin can be small and stubby or high and curved, and several small knuckles are visible on the dorsal ridge between the fin and the tail. Humpbacks often host barnacles and whale lice. Females are generally larger than males.

Fluke variations

RORQUALS

BLOW: The Humpback Whale makes a thick, orb-shaped blow that can reach up to 10 feet high and is visible at a great distance. From directly in front or behind, the blow may appear slightly heart-shaped.

OTHER DISPLAYS: An acrobatic whale, the Humpback dazzles whale watchers with high breaches that finish in a tremendous splash. Other behaviors that it may repeat several times include lob-tailing, flipper-slapping and spy-hopping. Humpbacks are often inquisitive, and they may approach boats if the boaters are non-harassing. When Humpbacks breathe and dive, they roll forward through the water and show a strongly arched back. The tail flukes are lifted high only on deep dives.

GROUP SIZE: Humpbacks commonly live in small groups of 2 or 3 members, but some groups may have 15 members, and Humpbacks are occasionally seen singly. Good feeding and breeding waters usually draw large groups.

FEEDING: The Humpback Whale feeds mainly by lunge-feeding or by bubble-netting, but much individual variation occurs in feeding styles. Its major foods are krill and schooling fish, such as anchovies and sardines.

REPRODUCTION: The courtship behavior of Humpbacks is elaborate and involves lengthy bouts of singing by the males. Mating usually occurs in warm waters, and single calves are born following a gestation period of about 11½ months. The calves stay close to their mothers and nurse for about 1 year. Females reach sexual maturity when they are about 40 feet long, and males when they are at least 38 feet long.

SIMILAR SPECIES: The North Atlantic Right Whale (p. 66) has callosities instead of knobs on its head, and it has much shorter flippers.

North Atlantic Right Whale

BALEEN WHALES

RIGHT WHALES

North Atlantic Right Whale

Eubalaena glacialis

Over the past few decades the right whales have been renamed and reclassified several times. Although some scientists still maintain there is only one species worldwide, there are currently three accepted species: the North Atlantic Right Whale, the North Pacific Right Whale (*E. japonica*) and the Southern Right Whale (*E. australis*). The highly endangered right whales were the most heavily pursued whales during the height of whaling. The name "right whale" is derived from mariners referring to it as the "right" whale to hunt: it is large, it has thick blubber and long baleen, and it swims slowly. The right whales were the first whales to be hunted commercially, and hundreds of thousands of right whales were killed to feed the demand for whale oil and other whale products. By 1920, the North Atlantic Right Whale was nearly extinct. Even today, there are fewer than 350 in the North Atlantic. The North Pacific Right Whale is even worse off; there may not even be 100 of them remaining. In the Southern Hemisphere, the numbers are better—as many as 12,000 Southern Right Whales still exist. The chances of seeing a North Atlantic Right Whale in certain areas of the northern east coast are excellent, especially in the Gulf of Maine and Cape Cod Bay. Spectacular research and viewing opportunities are found in calving waters off Georgia and Florida.

DIVE SEQUENCE

RIGHT WHALES

RANGE: found in polar and temperate waters of the North Atlantic; favors offshore waters rather than open ocean.

OTHER NAMES: Northern Right Whale, Black Right Whale, Biscayan Right Whale.

STATUS: critically endangered (USFWS).

ADULT LENGTH: up to 60 ft; average 45 ft.

ADULT WEIGHT: up to 95 tons; average 60 tons.

BIRTH LENGTH: 15–20 ft.

BIRTH WEIGHT: about 1 ton.

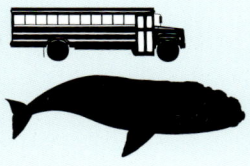

When North Atlantic Right Whales are seen in the wild, they can be quite approachable and inquisitive. Sometimes these whales are playful, bumping and pushing floating objects in the water. As a cautionary note, these qualities of inquisitiveness and playfulness can result in a whale playing near or with your boat. Some kayakers have been surprised when their boats were bumped or lifted in friendly jest.

As everyone knows, the greatest record-breaking animal in the world is the Blue Whale. The understated right whales, however, win a few titles of their own. Although it is not a very noble record to hold, right whales may well be the slowest whales on Earth. Less flattering still, these whales are one of the fattest whales, considering their weight per foot of body length. Their large tail is the broadest, relative to the length of the body. The one title these whales carry that may place them in the realm of nobility among cetaceans is for the largest testicles. With testes weighing in at 1100 pounds (each!), no other creature in the history of the world is as well-endowed. By comparison, the testes of the now-humbled Blue Whale weigh a mere 150 pounds.

RIGHT WHALES

DESCRIPTION: The robust North Atlantic Right Whale is easy to distinguish because of the callosities on its head and its lack of a dorsal fin. Distinct callosities grow on the "chin," on the rostrum, in front of the blowhole and above the eye. The body is nearly black, dark blue or dark brown, and it may show slight mottling or even patchy white spots on the belly. The mouth line is strongly arched upward and then drops nearly vertically to meet the eye. There is a distinct indentation behind the blowhole and then a bulge continuing over the back. The baleen plates can be as long as 8 feet. The flippers are large and spatula-shaped, and they are dark on both sides. At close range, the finger bones in a flipper can be seen as distinct ridges. The flukes, also dark both above and below, are smooth and pointed and have a noticeable notch in the middle. Females are generally larger than males.

BLOW: When viewed from the front or the rear, a right whale's distinctive blow is widely V-shaped. The sound is loud, and the spray can reach a height of 16 feet.

Callosities (front view)

RIGHT WHALES

shape of the baleen makes the mouth look like a cave. As the whale progresses through a concentration of copepods, the water enters the "cave" and passes through the baleen. All of the creatures are trapped against the silky hairs of the baleen plates and eventually swallowed. This whale skim-feeds wherever the food is present—at the surface or at depth.

REPRODUCTION: The right whales have never managed to recover their numbers the way other whales have. The reasons are not clear, but the combination of isolated populations and a slow reproductive rate may be responsible. Females give birth only once every 3 or 4 years. Although these whales court and mate year-round, peak conception appears to be in winter, as is calving, which indicates a gestation period of about 1 year. Both sexes are sexually mature when they are 75 to 80 percent of the adult length.

OTHER DISPLAYS: The North Atlantic Right Whale is quite acrobatic, considering its bulk. It often breaches, lob-tails, flipper-slaps and spy-hops. When it lob-tails, it may invert in the water to such a great extent that it appears to be doing a headstand.

GROUP SIZE: These right whales are typically found in small groups of 2 or 3 individuals. Sometimes a group numbering up to 12 members is seen. Seasonal feeding waters attract larger numbers of whales at certain times.

FEEDING: A remarkably efficient filter-feeder, the North Atlantic Right Whale feeds on some of the smallest zooplankton in the sea. The primary food is tiny krill and other copepods. To feed, the whale swims slowly with its mouth open. The unique

SIMILAR SPECIES: The Humpback Whale (p. 62) has knobs instead of callosities on its head and has much longer flippers.

Humpback Whale

Toothed Whales

Ocean Dolphin Family (Delphinidae)

This large family of toothed whales represents the true ocean dolphins, and it includes some of the most well-known cetaceans—aquariums, movies and anecdotal accounts have made Bottlenose Dolphins and Killer Whales world-famous cetaceans. There are at least 38 delphinid species worldwide, but some evidence indicates that certain species, like the Bottlenose Dolphin, may actually be two or three distinct species. At least 17 species of delphinids exist in east coast waters.

Although many people call the Killer Whale a whale, it is actually the largest dolphin in the world. The Killer Whale and its smaller relatives, such as the False Killer Whale and the pilot whales, are frequently referred to as "blackfish." Several authorities recognize these cetaceans as a separate family, but based on anatomical similarities, they are best classified as delphinids.

Porpoise Family (Phocoenidae)

Porpoises, which number only six species worldwide, are widely distributed in the world, and there is very little range overlap between species. Although they superficially resemble dolphins, porpoises have unique features that set them apart from the

Harbor Porpoise

better-known delphinids. They do not have a distinct beak, and their heads are quite rounded in appearance. Their body shape is a bit more robust than the streamlined dolphins, and their flippers are typically small and stubby. Unlike dolphins, which have conical teeth for holding and biting prey, porpoises have flattened, spade-shaped teeth that function to slice their prey.

Beaked Whale Family (Ziphiidae)

Beaked whales represent a large but poorly understood family of whales. They mainly inhabit deep ocean waters where people rarely visit, so accounts of their habits and behaviors are few. Some species of beaked whales have never been seen alive and are known only from individuals found dead on the shore or pulled in by fishing

TOOTHED WHALES

nets. There are 21 known species, but unrecorded beaked whale species may exist.

The ziphiids are referred to as beaked whales because all of them have a prominent snout. Males have two or four teeth that erupt on their lower jaw, either in the front or along the sides. Females and juveniles usually do not have erupted teeth of any sort. The lower jaws of the males and mature females may slope upward in the middle, giving the appearance that the "lip" is wrapping around the upper jaw. The teeth of the males further exaggerate this arch, and in some species the two teeth wrap over the upper jaw. This feature greatly reduces the ability of the whale to open its jaws, but beaked whales are believed to feed by sucking their prey into their mouth.

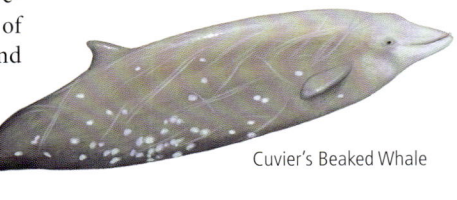
Cuvier's Beaked Whale

Dwarf Sperm Whale Family (Kogiidae) and Sperm Whale Family (Physeteridae)

Two families worldwide represent the sperm whales. The famous Sperm Whale, subject of Herman Melville's book *Moby Dick*, belongs to the family Physeteridae. The smaller sperm whales, in the family Kogiidae, are poorly understood, and very little information exists about their habits and natural history.

The most distinguishing feature of a sperm whale is its extremely large head. The largest of all, the head of a male Sperm Whale may be as much as one-third of the length of its body. The exact function of the large head is still under much debate. Some people think its serves to focus the echolocation clicks when the whale searches for food in very deep and dark water. Other theories suggest that the head, full of high-quality spermaceti oil, controls the buoyancy of the whale as it rises and dives through the water.

Pygmy Sperm Whale

71

Rough-toothed Dolphin
Steno bredanensis

DESCRIPTION: This streamlined dolphin has a distinctly conical head that helps differentiate it from similar dolphins. Its distinctive head and its large eyes give it an ancestral or primitive appearance. The beak is narrow, and it appears to have white "lips." Its overall body color is dark gray or bluish black, with slightly lighter sides and a white or pinkish white underside. The lighter sides create the appearance of a long, dark "cape" down its back. Along its sides and belly there may be distinct yellowish or pinkish blotches. The dorsal fin may be moderately to strongly falcate, and the flippers are large and pointed. The tail flukes are notched in the middle and have concave trailing edges. As this dolphin's name suggests, its teeth are rough; fine vertical ridges are present on all its teeth. Males are slightly larger than females.

BLOW: As with most dolphins, the blow is not readily visible, but it may be heard at short range as a rapid puffing sound.

OTHER DISPLAYS: This dolphin is a very fast swimmer, but it does not bow-ride as often as other species do. It may "porpoise" through the water, and it sometimes breaches. When breaching, the Rough-toothed Dolphin does not leave the water entirely, and it falls rather ungracefully back in. On rare occasions, these dolphins have been seen logging.

GROUP SIZE: This dolphin is most commonly seen

OCEAN DOLPHINS

RANGE: widespread but uncommon in warm waters of the world, normally where sea surface temperature is 77° F or above.

OTHER NAMES: Slopehead, Steno, Rough-toothed Porpoise, Black Porpoise.

STATUS: uncommon.

ADULT LENGTH: up to 8½ ft; average 7½ ft.

ADULT WEIGHT: up to 330 lb; average 275 lb.

BIRTH LENGTH: 3 ft.

BIRTH WEIGHT: unknown.

TOOTHED WHALES

in groups of 10 to 20 individuals, but on occasion single animals are seen, and at certain times several hundred may gather together.

FEEDING: The main food for this species is a variety of fish, squid and free-swimming octopus.

REPRODUCTION: Little is known about the reproduction of the Rough-toothed Dolphin, and different populations have varying patterns. The gestation period is unknown, but sexual maturity appears to occur at a little over 7 feet for both sexes.

SIMILAR SPECIES: The Bottlenose (p. 74), Spinner (p. 86) and Pantropical Spotted (p. 78) dolphins all have a distinct crease between the melon and the beak.

Bottlenose Dolphin

Bottlenose Dolphin
Tursiops truncatus

The familiar Bottlenose Dolphin is rivaled only by the Killer Whale as the most recognized cetacean species in the world. This fame is owed in part to aquarium performances and TV shows (such as *Flipper*), but also to the species' natural curiosity and friendliness toward humans. All over the world, the Bottlenose Dolphin is reported to hobnob with swimmers and perform acrobatics for admiring whale watchers. Of all cetaceans, this species is one of the best studied, especially in the fields of behavioral ecology, intelligence and communication.

The issue of non-human intelligence is loaded with misconceptions and biases. The main problem is that we have no guaranteed means of assessing intelligence in other animals. We can study their behavior and conduct performance tests, but there always seems to be a margin of doubt and error in every method we use. Nevertheless, it is clear that Bottlenose Dolphins are among the most intelligent of all mammals.

A close look at the brain of a dolphin reveals information about the potential intelligence of the creature. Second only to humans, a dolphin's brain is very large for its body size—a ratio called the "encephalization quotient"—and the only other creature to exhibit such a high ratio is the chimpanzee; on performance tests, however, Bottlenose Dolphins routinely figure problems out much faster than chimpanzees. When a test animal must activate switches to give correct answers, for example, chimpanzees routinely make random pokes at the switches before figuring out their purpose. Dolphins manage to trigger switches correctly with few, if any, random attempts. Another characteristic of a Bottlenose Dolphin's brain is the extremely well-developed cerebral cortex. The cerebral cortex is involved with the precise echolocation used by dolphins, but this part of the brain is also responsible for

OCEAN DOLPHINS

TOOTHED WHALES

RANGE: found around the world in tropical waters and in some temperate waters, but distribution is not continuous; favors bays and oceanic islands.

OTHER NAMES: Bottle-nosed Dolphin, Atlantic/Pacific Bottlenose Dolphin, Gray Porpoise, Gray Dolphin, Cowfish.

STATUS: common.

ADULT LENGTH: up to almost 13 ft; average 10 ft.

ADULT WEIGHT: up to 1430 lb; average 440 lb.

BIRTH LENGTH: about 3½ ft.

BIRTH WEIGHT: about 50 lb.

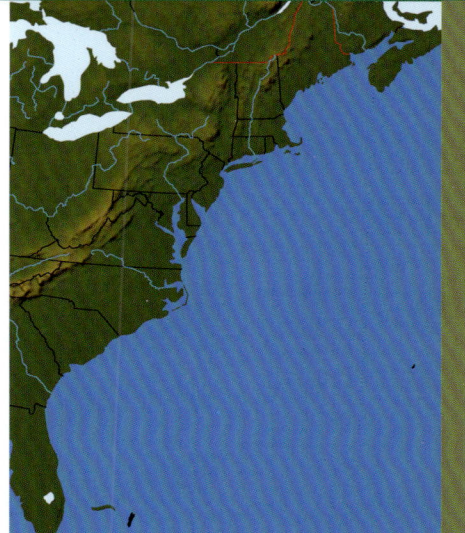

intellect, thought, language and memory.

Worldwide, three different subspecies of Bottlenose Dolphins have been described: *T.t. truncatus* in the Atlantic, *T.t. gilli* in the Pacific and *T.t. aduncus* in the Indian Ocean. Some researchers argue that they may even be distinct species. As well as subspecies, there appear to be two different varieties: the small and slender "inshore" dolphins, and the larger and more robust "offshore" dolphins. These two varieties exhibit social and behavioral differences. Offshore Bottlenose Dolphins appear to undergo slight seasonal migrations. Bottlenose Dolphins are year-round residents, and they can be found along most of the east coast, increasing in number toward Florida.

DESCRIPTION: This robust dolphin is mainly bluish gray, with subtle markings on the face, such as bands of darker gray from the eye to the flipper or from the eye to the top of the rostrum. The facial markings vary greatly between individuals and populations. Normally the undersides are light gray or whitish. The forehead is rounded, and there is a crease separating the short

OCEAN DOLPHINS

but prominent beak from the rest of the head. The dorsal fin is falcate and sometimes slightly hooked at the tip. This dolphin's flippers are dark colored on both sides, and they are broad at the base but taper to a long, slender point. The tail stock is thick, giving the dolphin a powerful appearance. Its flukes are distinctly notched, with swept-back tips. Members of offshore groups tend to be more robust than those of inshore groups.

BLOW: This dolphin does not make a distinct blow, but the sharp puffing sound is clearly audible at short range.

OTHER DISPLAYS: The Bottlenose Dolphin is famous for its acrobatics. It will frequently bow-ride and lob-tail, and when it breaches it often clears the water completely, sometimes with an added twist or somersault. Groups are highly active and play extensively at the surface.

OCEAN DOLPHINS

Inshore individuals rarely dive for longer than 5 minutes, but offshore dolphins may remain submerged for as long as 15 minutes.

GROUP SIZE: Group size can vary regionally and between populations. Offshore groups are larger, numbering up to 25 individuals, whereas inshore groups rarely number over 10. At times, up to 500 Bottlenose Dolphins have been sighted in open waters.

FEEDING: The Bottlenose Dolphin feeds on a surprising variety of creatures, such as fish, squid, octopus, eels, rays, shrimp, crabs and even small sharks. In some regions, this dolphin has been seen herding fish onto shore, following behind and grabbing the fish as they flop around on land. After consuming the fish, the dolphin slides or wriggles its way back into the water.

REPRODUCTION: Reproduction is well studied for this species. A female gives birth every 2 or 3 years. Gestation is 12 months, and nursing lasts 12 to 18 months. The age of first reproduction is not the same as the age of sexual maturity because social dynamics may influence mating. Sexual maturity may occur at 5 years (or later) for females, and at 10 or 11 years for males. The lifespan in the wild is 35 to 40 years.

SIMILAR SPECIES: An unspotted Atlantic Spotted Dolphin (p. 84) is very similar in shape but is noticeably smaller; the Spinner Dolphin (p. 86) has a streak between its eye and flipper; the Clymene Dolphin (p. 80) is smaller; the Rough-toothed Dolphin (p. 72) lacks a crease between the melon and the beak; the Risso's Dolphin (p. 98) has numerous scars over its body and does not have a prominent beak.

Atlantic Spotted Dolphin

TOOTHED WHALES

Pantropical Spotted Dolphin
Stenella attenuata

DESCRIPTION: This slender dolphin is best identified by its spots, but juveniles and a small percentage of adults lack spots altogether. It has a long beak and a subtle crease where the forehead slopes gently upward. The body is mainly bluish gray, with a dark "cape" over the back, light sides and a lighter belly. Most adults have dark spots on the light areas and light spots on the dark areas. Streaks are present on the face, and there is usually a dark stripe from the eye to the flipper. The flippers are dark but often have faint light spots, as does the falcate dorsal fin. The flukes are slightly notched in the middle and are triangular when viewed from above, without the swept-back appearance of most dolphin species. The coastal variety of this species tends to be more robust than the offshore variety.

BLOW: Most of these dolphins do not make a visible blow. Instead, listen for a sharp puffing sound.

OTHER DISPLAYS: This dolphin is very acrobatic. It readily bow-rides, lob-tails and breaches, and it occasionally performs a somersault before it splashes back in. When a herd of these dolphins moves quickly, the frothy water is visible at quite a distance. This dolphin is a fast swimmer, making long, shallow leaps.

GROUP SIZE: This species is rarely seen singly; groups number anywhere from 5 dolphins to 3000. The inshore variety is rarely seen in groups larger than 100.

FEEDING: These dolphins tend to associate with

OCEAN DOLPHINS

RANGE: found in tropical waters and certain warm-temperate waters of the Atlantic, Pacific and Indian oceans.

OTHER NAMES: Spotted Dolphin, Bridled Dolphin, Slender-beaked Dolphin, White-spotted Dolphin.

STATUS: common.

ADULT LENGTH: up to 8½ ft; average 6½ ft.

ADULT WEIGHT: up to 310 lb; average 220 lb.

BIRTH LENGTH: 33 in.

BIRTH WEIGHT: unknown.

REPRODUCTION: Mating and calving occur at any time of the year. Females give birth every 2 to 4 years. Gestation takes just over 11 months. Males reach sexual maturity when they are about 6 feet in length, or about 6 to 8 years old. Females are sexually mature between the ages of 4 and 8, when they are just over 6 feet long.

Spinner Dolphins (p. 86), and to limit competition, the two feed on different prey. Pantropical Spotted Dolphins feed mainly on surface (epipelagic) fish (especially flying fish), squid and some crustaceans, whereas Spinner Dolphins feed on similar prey but at a greater depth. In addition, the two species feed at different times.

SIMILAR SPECIES: The Atlantic Spotted Dolphin (p. 84) is more robust and usually more heavily spotted; the Spinner Dolphin (p. 86) lacks spots; the Bottlenose Dolphin (p. 74) is larger, unspotted and more robust.

Clymene Dolphin
Stenella clymene

DESCRIPTION: The Clymene Dolphin was identified as a unique species and separated from the Spinner Dolphin (p. 86) in 1981. Its body is varying shades of gray, with a distinct tricolored pattern on its sides: dark gray above, a streak of lighter gray through the middle, and pale gray along the belly. It has a fairly narrow beak and dark "lips," giving it a mustached appearance. Often the top side of the beak has a dark line from the tip to the melon. There is a distinct crease between the beak and the melon. The dorsal fin is slightly falcate and quite large relative to the body size.

BLOW: In general, dolphins do not make visible blows, but the sharp puffing sound of their exhalation and inhalation can be heard at close range.

OTHER DISPLAYS: An acrobatic species, these dolphins breach the water in high jumps and may spin during the jumps. They frequently bow-ride against large boats. Clymene Dolphins often play and associate with Spinner Dolphins and Short-beaked Saddleback Dolphins (p. 88).

GROUP SIZE: Typical group size is 60 to 80 members, but single individuals have been seen, and groups as large as 1000 have been known to form. Such a large group is temporary, and the reason for the gathering is not known.

FEEDING: These dolphins live in deep water, and they appear to feed on small, deepwater fish

OCEAN DOLPHINS

TOOTHED WHALES

RANGE: found in deep, warm equatorial waters of the Atlantic, from New Jersey to Brazil.

OTHER NAMES: Short-snouted Spinner Dolphin, Atlantic Spinner Dolphin, Helmet Dolphin.

STATUS: unknown.

ADULT LENGTH: up to 6½ ft; average 6¼ ft.

ADULT WEIGHT: up to 200 lb; average 180 lb.

BIRTH LENGTH: about 2½ ft.

BIRTH WEIGHT: about 22 lb.

that are known to rise to the surface at night during vertical migration. As nighttime feeders, Clymene Dolphins reduce competition for the prey species available during the day. Some small squid species are also consumed.

REPRODUCTION: Few details are known of the reproduction of Clymene Dolphins. As a deepwater species, their activities are difficult to study. Like other *Stenella* species, a female will give birth to 1 calf after a gestation of about 11 months. Females probably give birth once every 3 years, as nursing of a calf can continue for as long as 2 years in *Stenella* species.

SIMILAR SPECIES: The Spinner Dolphin (p. 86) is lighter and usually has a well-defined streak between the eye and the flipper; the Striped Dolphin (p. 82) has a distinct stripe from the eye to under the tail stock and a zig-zag pattern on its sides.

Spinner Dolphin

OCEAN DOLPHINS

Striped Dolphin
Stenella coeruleoalba

DESCRIPTION: This streamlined dolphin is easy to recognize because of the distinct thin stripe running from its eye to the underside of its tail stock. The Striped Dolphin is gray over most of its back, light gray on its sides and white or pinkish underneath. The border between the dark and light gray colors forms a zigzag below the dorsal fin. The dorsal fin is curved, and both the fin and flippers are dark on both sides. The pale gray flukes are slightly concave and notched in the middle. The beak is prominent and dark colored.

BLOW: This dolphin does not have a visible blow, but if you are close enough, the sound of its breathing is audible.

OTHER DISPLAYS: The Striped Dolphin is highly acrobatic, and it breaches the water to heights up to 23 feet. It may turn somersaults in the air or turn tail spins, re-entering the water in a graceful dive. This dolphin can "porpoise" through the water upside-down.

GROUP SIZE: Striped Dolphins are commonly found in herds of 100 to 500 members. At times, up to 3000 have been seen together.

FEEDING: This dolphin's primary diet is small to medium-sized fish and squid.

REPRODUCTION: Successful mating can occur in either summer or winter. After a gestation period of 12 to 13 months, the female gives birth to 1 calf. Females are sexually mature at about 90 percent of their adult length. Males, though they are sexually

OCEAN DOLPHINS

RANGE: found in warm-temperate, subtropical and tropical waters around the world.
STATUS: insufficient data worldwide; locally common in some areas.
ADULT LENGTH: up to 9 ft; average 7 ft.
ADULT WEIGHT: up to 350 lb; average 260 lb.
BIRTH LENGTH: 3–3½ ft.
BIRTH WEIGHT: unknown.

TOOTHED WHALES

mature when they are about 9 years old, are not socially mature enough to mate until they are 16 years old.

SIMILAR SPECIES: No other dolphin in the region has a stripe from the eye to under the tail stock. The Pantropical Spotted (p. 78) and Atlantic Spotted (p. 84) dolphins have spots; the Clymene (p. 80) and Spinner (p. 86) dolphins have less-defined coloration.

Clymene Dolphin

Pantropical Spotted Dolphin

OCEAN DOLPHINS

Atlantic Spotted Dolphin

Stenella frontalis

DESCRIPTION: The key feature of this dolphin is its spots, which become more numerous as the dolphin gets older. Juvenile dolphins and the odd adult lack spots altogether. Unspotted adults are more common in offshore populations. Without spots, this dolphin closely resembles a small Bottlenose Dolphin (p. 74). On the light gray belly the spots are dark, and on the darker sides the spots are light. There is a dark gray "cape" over the back and dorsal fin, and sometimes there is a faint zigzag below the dorsal fin where this dark gray meets the lighter gray of the sides. There is a distinct crease separating the melon from the moderately long beak. The dorsal fin is tall and falcate.

BLOW: Like most other dolphins, the blow is audible instead of visible. The sharp puffing sound carries well in cool air.

OTHER DISPLAYS: This species is very acrobatic, and it breaches in high jumps out of the water. It swims very fast and often approaches passing boats to bow-ride.

GROUP SIZE: Typical pod size is only 5 to 10 individuals, especially in near-shore groups. Like many species of dolphins, their social relations seem fluid, and small pods regularly

84

OCEAN DOLPHINS

RANGE: found along shelf waters of the Atlantic, from Massachusetts to Brazil.
OTHER NAMES: Gulf Stream Spotted Dolphin, Cuvier's Dolphin, Long-snouted Dolphin.
STATUS: probably stable.
ADULT LENGTH: up to 7½ ft; average 6½ ft.
ADULT WEIGHT: up to 315 lb; average 270 lb.
BIRTH LENGTH: about 3 ft.
BIRTH WEIGHT: unknown.

mix with others in groups of 50 individuals or more. Additionally, they mix with other species such as Bottlenose and Spinner (p. 86) dolphins.

are sexually mature when they are about 8 or 9 years old, and males when they are 12 to 15 years old. Social maturity may influence their age of first mating. On average, females give birth to 1 calf every 3 years. Gestation is 11 to 12 months.

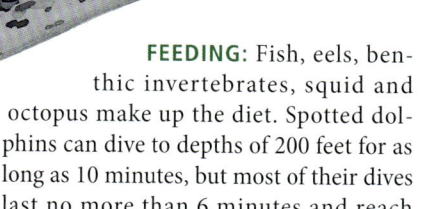

FEEDING: Fish, eels, benthic invertebrates, squid and octopus make up the diet. Spotted dolphins can dive to depths of 200 feet for as long as 10 minutes, but most of their dives last no more than 6 minutes and reach depths of only about 30 feet.

REPRODUCTION: Sexual maturity occurs a different age for the two sexes. Females

SIMILAR SPECIES:
The Pantropical Spotted Dolphin (p. 78) is sleaker and usually has fewer spots; the Spinner Dolphin (p. 86) lacks spots; the Bottlenose Dolphin (p. 74) is larger and lacks spots.

Pantropical Spotted Dolphin

OCEAN DOLPHINS

Spinner Dolphin
Stenella longirostris

DESCRIPTION: This streamlined dolphin gets its slender appearance from its long beak and swept-back tail. Its overall color is gray, and it is generally dark on the back, lighter on the sides and whitish on the belly. Obvious markings may include some streaking on the face and between the eye and the flipper, and a creamy or pinkish patch on the belly in line with the dorsal fin. The flukes are strongly swept back and only mildly notched in the middle. The dorsal fin is tall and erect, sometimes with a concave curve on the leading edge. The forehead slopes gently upward from the slight crease that separates it from the beak.

BLOW: Like most dolphins, the Spinner Dolphin does not make a blow. Its breathing sound can be heard, however, and when several dolphins are traveling fast at the surface, they make sprays that can resemble a blow.

OTHER DISPLAYS: One of the most acrobatic of all cetaceans, the Spinner Dolphin is famous for its spinning breaches. The dolphin launches itself up to nearly 10 feet clear of the water and spins along its longitudinal axis. It rarely breaches without spinning. No other dolphin spins as often or as well as this dolphin does; other acrobatic dolphins will twist only somewhat or perform somersaults during a breach. Spinner Dolphins also readily bow-ride and play.

GROUP SIZE: Group size is variable, and a group will readily mingle with nearby groups (and sometimes with other species, such as the Pantropical Spotted Dolphin, p. 78). Anywhere from 1 to 1000 dolphins may be seen at a time.

OCEAN DOLPHINS

TOOTHED WHALES

RANGE: found in tropical and subtropical waters of the Atlantic, Pacific and Indian oceans.

OTHER NAMES: Long-snouted Spinner Dolphin, Long-snouted Dolphin, Rollover, Long-beaked Porpoise.

STATUS: common.

ADULT LENGTH: up to 7 ft; average 5½ ft.

ADULT WEIGHT: up to 165 lb; average 130 lb.

BIRTH LENGTH: 2½ ft.

BIRTH WEIGHT: unknown.

FEEDING: This dolphin feeds mainly on fish, small squid and shrimp found well below the surface (the mesopelagic region), thus limiting competition for food with the Pantropical Spotted Dolphin, which feeds more on surface (epipelagic) prey.

REPRODUCTION: Females may give birth every 2 years, and gestation is just over 10 months. Males are sexually mature when they reach 5½ feet in length, at anywhere from 6 to 11 years old. Females mature earlier, at a length of just under 5½ feet, or about 4 or 5 years old.

SIMILAR SPECIES: The Clymene Dolphin (p. 80) is usually darker in color and has a faint or absent streak between the eye and flipper; the Bottlenose Dolphin (p. 74) is larger and more robust; the Rough-toothed Dolphin (p. 72) lacks a crease between the melon and the beak.

Clymene Dolphin

Short-beaked Saddleback Dolphin
Delphinus delphis

More so than any other dolphin group, the saddleback dolphins seem to confound both taxonomists and biologists alike. At least 20 species have been proposed worldwide, as well as several subspecies. Officially, the group was recently split into two species: the Short-beaked Saddleback Dolphin (*D. delphis*) and the Long-beaked Saddleback Dolphin (*D. capensis*). As the names suggest, the latter has a longer beak than the former, and, in general, the Long-beaked Saddleback Dolphin has a longer and slightly slimmer body, more muted coloration and a thicker dark line from the lower jaw to the flipper than the Short-beaked Saddleback Dolphin. Although both species occur on the west coast of North America, only the Short-beaked is known to occur in the waters of the east coast. Mysteriously, it has not been recorded in Florida waters since the 1960s, and no one knows the reason for this odd disappearance.

Where they occur, Short-beaked Saddleback Dolphins are usually a very abundant species. They are also very sociable and are regularly seen with both Striped Dolphins (p. 82) and Risso's Dolphins (p. 98).

Like many other dolphins, saddleback dolphins are able to make an array of different sounds: they click, squeak and whistle, and they have a particularly high-pitched squeal that can even be heard from

OCEAN DOLPHINS

RANGE: widely distributed in warm-temperate, subtropical and tropical waters of the world.

OTHER NAMES: Short-beaked Common Dolphin, Common Dolphin, Criss-cross Dolphin.

STATUS: common.

ADULT LENGTH: up to 8½ ft; average 6½ ft.

ADULT WEIGHT: up to 300 lb; average 170 lb.

BIRTH LENGTH: about 2½ ft.

BIRTH WEIGHT: unknown.

directly away from the threat. Even when they are resting, they are alert for danger. Dolphins literally have the ability to sleep in halves: they are able to rest one eye while the other stays alert. Their brain also rests one hemisphere at a time, allowing for a constant state of awareness.

above the water. When large groups of saddleback dolphins are playing, they can be so loud and loquacious that they are heard long before they are seen. If the herd is startled, they bunch tightly together for safety.

Saddleback dolphins are very wary of Killer Whales (p. 110) and sharks, and if the alarm is called they flee at full speed

Unfortunately, one danger to which dolphins are oblivious is fishing nets, and a significant number of saddleback dolphins are killed each year by fishing activities. Saddleback dolphins are at high risk of getting caught in nets because in some waters they associate closely with yellowfin tuna. Only Pantropical Spotted Dolphins (p. 78) and Spinner Dolphins (p. 86) incur greater losses from drowning in nets.

OCEAN DOLPHINS

DESCRIPTION: This small to mid-sized dolphin is easy to identify because of its unique coloration. On each side, beginning behind the eye, there is an arching, tan-colored patch. This patch can be quite yellowish, but on some animals it is a dull muddy gray. The tail stock has a light gray area, also in an arch, and where the gray meets the yellowish patch on the side, there is a distinct criss-crossing pattern. The back is mainly dark brown or bluish black, and the undersides are whitish. The beak is dark colored and sometimes white-tipped, and there are streaks of white or gray around the face. A dark circle always surrounds the eye, and a dark streak runs from the lower jaw to the flipper. The broad flippers are dark on both

Long-beaked
Saddleback Dolphin

Short-beaked
Saddleback Dolphin

OCEAN DOLPHINS

GROUP SIZE: Perhaps the most numerous dolphins, the saddleback dolphins have a world population that may well be in the millions. Group size is extremely variable, from a few dozen individuals to over 10,000. Along the east coast, group size tends to be fewer than 100 dolphins, but groups of up to 3000 have formed in good feeding waters, especially around New England.

FEEDING: Saddleback dolphins feed mainly on small schooling fish, such as herring, anchovies and sardines. A group of dolphins typically herds the fish into a tight ball and then begins feeding. Such cooperative strategies are used by many species of dolphins. Feeding dolphins can attract seabirds to an area because the birds benefit from the herded fish. Some people speculate that the reverse may be true too; a flock of feeding seabirds indicates the whereabouts of fish and might attract dolphins to the area.

sides, as are the pointed flukes. The triangular dorsal fin is centrally located, and often the center is a lighter gray than the rim. The flippers, dorsal fin and flukes all trail backward, giving the dolphin a streamlined appearance.

BLOW: The blow of a saddleback dolphin is not visible. At close range, however, the sound of its breathing can be heard when it is in mid-leap.

OTHER DISPLAYS: Saddleback dolphins are exceptional acrobats, and they frolic and play in noisy displays. They may leap high out of the water and fall gracefully back in with little splash. They frequently bow-ride, and other displays include flipper slaps, jaw slaps and lob-tailing.

REPRODUCTION: The reproductive ages and lengths of saddleback dolphins can vary between different regions and subspecies, but generally females are sexually mature when they are 6 feet long, and males when they are 6½ feet long. Mating and calving can occur at any time of the year, and gestation takes 11 to 12 months. The social structure of a group might influence courtship and mating.

SIMILAR SPECIES: The Striped Dolphin (p. 82) has a stripe from its eye to under its tail stock; the Atlantic White-sided Dolphin (p. 92) has a tricoloured body and a bicoloured beak; the White-beaked Dolphin (p. 96) has an entirely white beak.

OCEAN DOLPHINS

Atlantic White-sided Dolphin
Lagenorhynchus acutus

This large and colorful dolphin inhabits cool to temperate waters in the North Atlantic. Common around Cape Cod and the Gulf of Maine, the Atlantic White-sided Dolphin can be seen in pods of a couple of dozen individuals. Large pods, numbering about 60, are occasionally encountered. Sometimes pods numbering several hundred dolphins are reported. For unknown reasons, this dolphin seems to strand more frequently than others in the region. Sometimes individuals strand, and sometimes groups of several dozen or even a hundred or more are reported stranded. These strandings may occur at any time of the year, and despite the efforts of researchers and volunteers, few of the dolphins are saved.

Another common dolphin in the same region is the Short-beaked Saddleback Dolphin (p. 88), and these two species are often misidentified. The Short-beaked Saddleback Dolphin has the yellow side streak to the front of the dorsal fin, rather than to the rear. Both species are playful and quick to bow-ride. Atlantic White-sided Dolphins are sometimes even seen riding the bow waves of large whales. These dolphins, and other members of the same genus, are often

OCEAN DOLPHINS

RANGE: found in subarctic waters of the North Atlantic: western edge of range extends from Greenland to Maryland; eastern edge stretches from western Norway to Great Britain.

STATUS: low population, but stable.

ADULT LENGTH: up to 9½ ft; average 8 ft.

ADULT WEIGHT: up to 500 lb; average 450 lb.

BIRTH LENGTH: about 3½ ft.

BIRTH WEIGHT: about 45 lb.

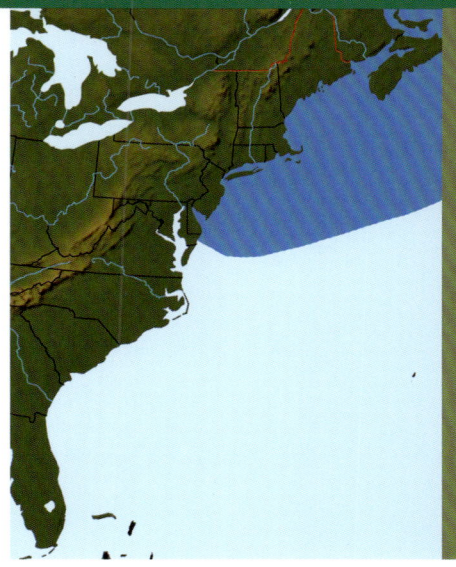

TOOTHED WHALES

Atlantic White-sided Dolphins tend to prefer cool, low salinity waters. Although these dolphins are not strongly migratory, there is movement between inshore and offshore waters in response to food availability. Normally, these dolphins are considered an open ocean species, but they come to near shore waters in spring in search of mackerel.

The total population of Atlantic White-sided Dolphins is only about 300,000 individuals. Fewer than 1000 of these dolphins are hunted every year, but many more accidental deaths from fishing nets occur every year.

referred to as "lags," a diminutive of their scientific name. As a group, these dolphins are not only acrobatic but also very sociable. Together they surf ocean waves, catch wakes, ride bow waves and "porpoise" in unison.

OCEAN DOLPHINS

DESCRIPTION: The Atlantic White-sided Dolphin actually appears to be tricolored; the belly and lower sides are white, the sides are gray, and the back is dark gray or black. Its facial markings are quite distinct; it has a black eye ring and a bicolored beak, black above and gray below. A faint gray stripe may connect the leading edge of the flipper with the eye ring or rear margin of the lower jaw. There is a diagnostic lateral yellow stripe to the rear of the dorsal fin, sometimes with a leading portion of white. The fins are all strongly pointed; the flippers are about 12 inches in length, the dorsal fin may be up to 19 inches in height, and the tail flukes are up to 24 inches across. Females are a bit smaller than males.

BLOW: There is no discernible blow, but they often clear the water every 10 to 15 seconds to breathe. Sometimes they swim just below the surface, so their head cuts the surface of the water to make a spray that is often mistaken for a blow.

OCEAN DOLPHINS

When they are wary or moving slowly they will rise just enough to expose their blowhole and then curl back under.

OTHER DISPLAYS: Playful and acrobatic, Atlantic White-sided Dolphins are known to breach high out of the water and bow-ride with fast boats. They often swim just below the surface of the water, creating a spray off their head and dorsal fin.

GROUP SIZE: Pods of dolphins numbering several dozen are most common. Large pods of several hundred sometimes form, especially if they are following large whales that are feeding. Social dynamics also influence pod segregation. There is a strong tendency for young, sexually immature dolphins to form their own pod, much like teenagers hanging out with each other.

FEEDING: Lanternfish, herring, mackerel, squid, smelt, hake and shrimp appear to make up the bulk of the diet. Atlantic White-sided Dolphins are not considered deep divers; although they can be found in waters up to 1000 feet deep, they do not dive to the bottom and rarely stay underwater for more than 4 minutes. Sometimes they cooperate in their efforts to surround and attack schools of fish.

REPRODUCTION: Females give birth to 1 calf every 2 years. The peak calving period is in June and July but extends from May to August. Females mate again when the calf is about 1 year old, and the gestation is about 11 months. Calves are weaned when they are 1½ years old. Sexual maturity is reached after 6 years of age. These dolphins usually live 25 to 30 years.

SIMILAR SPECIES: The Short-beaked Saddleback Dolphin (p. 88) has a yellow streak on its side behind its eye and a distinct criss-crossing pattern on its side below the dorsal fin; the White-beaked Dolphin (p. 96) has an entirely white beak; the Striped Dolphin (p. 82) has a stripe from its eye to under its tail stock.

Short-beaked Saddleback Dolphin

OCEAN DOLPHINS

White-beaked Dolphin
Lagenorhynchus albirostris

DESCRIPTION: This robust dolphin is dark gray or nearly black, with indistinct white patches on the beak, sides, belly and back behind the dorsal fin. There is considerable variation in the coloration between individuals. The belly, "chin" and beak are white. The melon is dark gray or black, and the eye is in the dark coloration. The white beak contrasts sharply with the dark melon. The dorsal fin is long, narrow and strongly falcate on the trailing edge.

BLOW: White-beaked Dolphins make a small, slightly splashy blow as though they begin exhaling before their blowhole is clear of the water. The sound of their blow is audible at close range.

OTHER DISPLAYS: These dolphins are fond of bow-riding, and they frequently make high jumps out of the water. White-beaked Dolphins are known to accompany Fin Whales (p. 60) and Humpback Whales (p. 62) to good feeding areas.

GROUP SIZE: Group size varies from 5 to 50 members, though about 30 is the most common. Like other dolphin species, large gatherings of 1000 individuals or more sometimes form. The reason for these large gatherings is not known. These dolphins are sometimes encountered in mixed groups with Bottlenose (p. 74) and Atlantic White-sided (p. 92) dolphins.

OCEAN DOLPHINS

RANGE: found in the North Atlantic, from the Barents Sea to Cape Cod.
STATUS: stable.
ADULT LENGTH: up to 10½ ft; average 9½ ft.
ADULT WEIGHT: up to 350 lb; average 290 lb.
BIRTH LENGTH: about 4 ft.
BIRTH WEIGHT: about 90 lb.

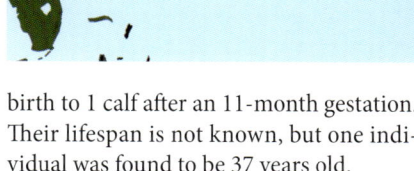

FEEDING: White-beaked Dolphins are found in shelf and offshore waters, but they seem to prefer water that is no more than 650 feet deep. Their primary food is schooling fish such as haddock, cod, hake and whiting. Squid and crustaceans, primarily crabs, are also eaten. Seasonal fluctuations in their preferred prey species influence their inshore/offshore movement.

REPRODUCTION: White-beaked Dolphins become physically and socially mature enough to reproduce somewhere between the ages of 7 and 12 years old. Mating and calving occur in summer. Females give birth to 1 calf after an 11-month gestation. Their lifespan is not known, but one individual was found to be 37 years old.

SIMILAR SPECIES: The Atlantic White-sided Dolphin (p. 92) has a tricoloured body and a bicoloured beak; the Short-beaked Saddleback Dolphin (p. 88) has a yellowish patch behind its eye; the Striped Dolphin (p. 82) has a stripe from its eye to under its tail stock.

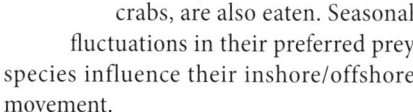

Atlantic White-sided Dolphin

Risso's Dolphin
Grampus griseus

DESCRIPTION: Unlike almost every other dolphin or whale, the Risso's Dolphin becomes so heavily scarred that old individuals can appear almost white. The scars result from teeth scratches incurred during sparring matches with each other or from the hooks on the tentacles of squid. This dolphin has a stout body and a blunt head. It has an unmistakably tall dorsal fin, long, sickle-shaped flippers and pointed, trailing flukes. Males are generally larger than females.

BLOW: The Risso's Dolphin does not make a visible blow, but the sound of its breathing is audible if the animal is close.

OTHER DISPLAYS: These dolphins are not as acrobatic as the smaller dolphins, but younger animals may breach fully out of the water. They are also seen spy-hopping, lob-tailing and flipper-slapping. Although Risso's Dolphins rarely bow-ride, they often surf in waves and wakes.

GROUP SIZE: Risso's Dolphins are regularly seen in groups of 3 to 50 individuals. Some groups can be as large as 1000, especially in food-rich waters. These dolphins sometimes associate with Atlantic White-sided Dolphins (p. 92).

OCEAN DOLPHINS

RANGE: found in deep tropical and temperate waters throughout the world; avoids cold polar waters.
STATUS: common.
ADULT LENGTH: up to 13 ft; average 10 ft.
ADULT WEIGHT: up to 1100 lb; average 880 lb.
BIRTH LENGTH: 4–5 ft.
BIRTH WEIGHT: unknown.

TOOTHED WHALES

are sexually mature when they are about 8½ feet long.

SIMILAR SPECIES: No other similar-sized cetacean shows such heavy scarring. The Bottlenose Dolphin (p. 74) has a more distinct beak; the Melon-headed Whale (p. 100) is smaller; the Pygmy Sperm Whale (p. 130) and the Dwarf Sperm Whale (p. 132) have underslung lower jaws.

FEEDING: In deep water, these agile dolphins feed primarily on squid, though some medium-sized fish are also eaten.

REPRODUCTION: Mating occurs primarily in spring, though the exact timing varies among different populations. Females give birth to 1 calf, usually in summer, after a gestation of 13 to 14 months. Both sexes

Bottlenose Dolphin

Melon-headed Whale
Peponocephala electra

DESCRIPTION: This delphinid is primarily dark gray or bluish black. The face and head are slightly darker than the main part of the body, and there is a similar darker saddle patch around the dorsal fin. The anchor patch on its underside is usually grayish in color; it is never pure white. Its head is conical, and the upper line of the head appears quite rounded. The distinctly upturned mouth often exhibits white or pinkish "lips" as a result of scarring. This whale's flippers are quite long and taper gently to a point, and the dorsal fin is tall and falcate. Its flukes are broad, with a concave trailing edge and pointed tips. Females average slightly larger than males.

BLOW: This whale does not make a distinct blow, but as with most delphinids, the sharp puffing sound can be heard across the water. When it travels, this whale can make foamy splashes that resemble a blow.

OTHER DISPLAYS: When startled, this whale swims very fast, up to 17 mph. It is usually wary of boats, but it has been known to bow-ride for short periods. It infrequently spy-hops and may breach clear of the water. In general, this whale is considered sociable and quite active.

GROUP SIZE: Although Melon-headed Whales may be seen singly, they are usually found in pods of 10 to 30 individuals and occasionally in pods of 100 to 200. At times, gatherings of up to 2000 have been recorded. These whales are known to mingle with dolphins such as Spinner Dolphins (p. 86).

OCEAN DOLPHINS

TOOTHED WHALES

RANGE: primarily found in warm tropical and subtropical waters around the world; most occurrences in temperate waters are probably associated with warm currents.

OTHER NAMES: Many-toothed Blackfish, Little Killer Whale, Electra Dolphin.

STATUS: probably stable.

ADULT LENGTH: up to 9 ft; average 7 ft.

ADULT WEIGHT: up to 440 lb; average 350 lb.

BIRTH LENGTH: 3¼ ft.

BIRTH WEIGHT: 35 lb.

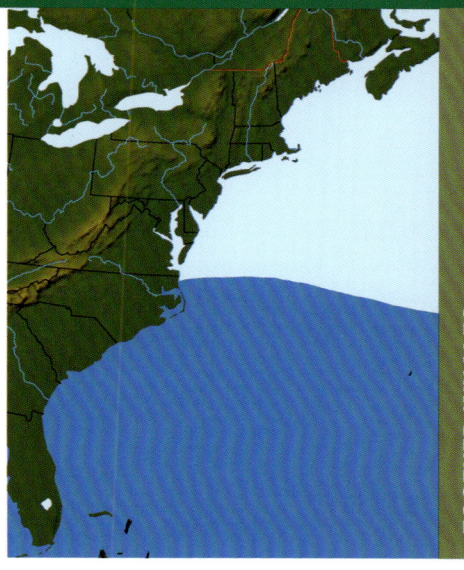

probably favor spring and summer months. Both sexes are sexually mature at or just beyond 7 feet.

FEEDING: Melon-headed Whales are fast and agile swimmers, and they feed on a variety of schooling fish, octopus, squid and some large shrimp.

REPRODUCTION: Very little is known about the reproduction of this species, but gestation appears to take about 12 months, and peak calving and mating periods

SIMILAR SPECIES: The Pygmy Killer Whale (p. 102) has flippers with rounded tips; the False Killer Whale (p. 104) is much larger; the Risso's Dolphin (p. 98) is heavily scarred.

Pygmy Killer Whale

OCEAN DOLPHINS

Pygmy Killer Whale
Feresa attenuata

DESCRIPTION: People rarely get close enough to this little bluish black whale to see the details in its color pattern. There is a dark gray or black "cape" over the face and back, the sides are a lighter gray, and there is a whitish patch on the belly in line with the dorsal fin. Between the flippers there is a pale gray anchor patch. The "lips" are white, and sometimes the "chin" is as well. This whale's flippers are long and dark, with rounded tips. The dorsal fin is tall, and the somewhat curved trailing edge is wavy on some individuals. The flukes are swept back, with pointed tips and a small notch in the middle. Males are larger than females.

BLOW: This whale does not make a visible blow, but a puffing sound may be heard. Occasionally this whale makes an audible growling sound at the surface.

OTHER DISPLAYS: Pygmy Killer Whales are considered to be quite wary of boats, but there are infrequent reports of bow-riding. Other activities that might be seen are logging, spy-hopping, lob-tailing and breaching, but overall this species is not acrobatic. For unknown reasons, these whales often strand.

OCEAN DOLPHINS

RANGE: inhabits tropical waters and some subtropical waters around the world; not abundant anywhere within its range.

OTHER NAMES: Slender Blackfish, Slender Pilot Whale.

STATUS: rare, but probably stable.

ADULT LENGTH: up to 9 ft; average 7 ft.

ADULT WEIGHT: up to 500 lb; average 350 lb.

BIRTH LENGTH: about 32 in.

BIRTH WEIGHT: unknown.

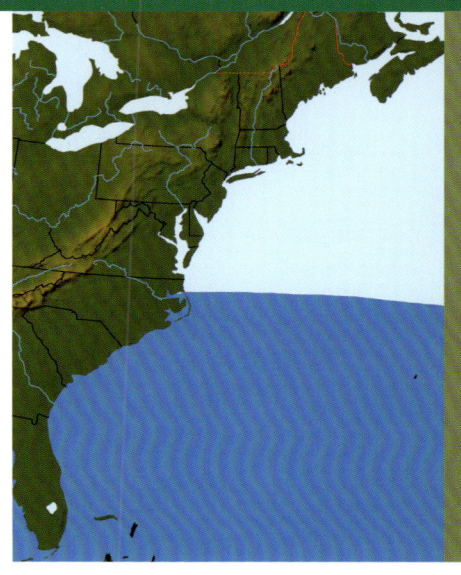

GROUP SIZE: This whale is most likely to be seen in a group of 10 to 35 individuals, but single animals are sometimes encountered. At certain times, several hundred have been seen together.

FEEDING: This whale's diet appears to be mainly squid, octopus and fish. It is also known to feed on certain other marine mammals, such as small dolphins or porpoises.

REPRODUCTION: Little is known about the reproduction of this species, but it appears to breed and give birth in spring and summer. The gestation period is unknown, but it is probably similar to that of other whales of its size, roughly 11 months. Sexual maturity appears to be at about 7 feet in length for both sexes.

SIMILAR SPECIES: The Melon-headed Whale (p. 100) has pointed flippers; the False Killer Whale (p. 104) is much larger; the Risso's Dolphin (p. 98) is heavily scarred.

Melon-headed Whale

False Killer Whale
Pseudorca crassidens

DESCRIPTION: The False Killer Whale is almost entirely black or deep gray, except for a light gray W- or anchor-shaped patch on its underside, between the flippers. The snout is blunt and rounded, with the upper jaw protruding past the lower jaw. It has up to 24 teeth in the upper jaw and the same in the lower jaw. The dorsal fin is upright, slightly rounded and located in the center of the dorsal line. The flippers appear slightly S-shaped because of the humped leading edge. The tail flukes are mildly pointed and notched in the middle. Males are generally larger than females.

BLOW: Like most small whales, the False Killer Whale makes no visible blow.

OTHER DISPLAYS: When this whale surfaces, it often rises out of the water, or "porpoises," exposing its flippers. Occasionally, it has its mouth open, showing its teeth. The False Killer Whale breaches frequently and twists back into the water with a tremendous splash. It also bow-rides, wake-rides and even makes acrobatic leaps clear out of the water.

GROUP SIZE: False Killer Whales are commonly seen in groups of 10 to 50 individuals. Gatherings of several hundred may occur throughout the year.

OCEAN DOLPHINS

RANGE: found in open ocean in tropical and warm-temperate waters.
OTHER NAMES: Pseudorca, False Pilot Whale, Blackfish.
STATUS: rare.
ADULT LENGTH: up to 20 ft; average 17 ft.
ADULT WEIGHT: up to 2.2 tons; average 1.6 tons.
BIRTH LENGTH: 5–6½ ft.
BIRTH WEIGHT: 180 lb.

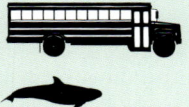

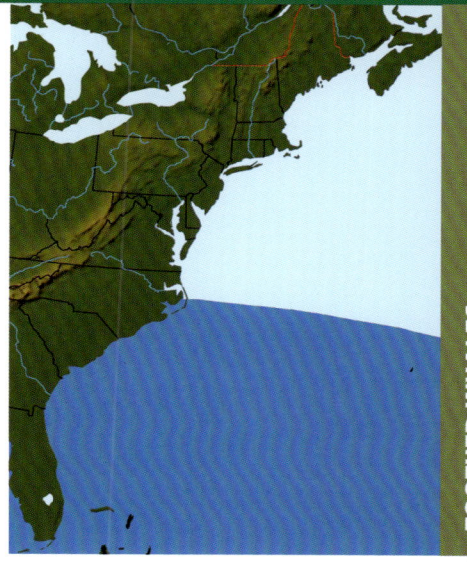

TOOTHED WHALES

gestation period is believed to be between 14 and 16 months. Males and females mature at roughly 80 percent of their adult body length.

SIMILAR SPECIES: The Short-finned Pilot Whale (p. 106) has a more bulbous melon and

FEEDING: These fast whales feed mainly on squid and large fish, such as yellowfin tuna. They have been known to attack and kill other cetaceans, such as dolphins and even, on at least one recorded occasion, a Humpback Whale (p. 62).

a falcate dorsal fin; the Pygmy Killer Whale (p. 102) is much smaller; the Melon-headed Whale (p. 100) is much smaller; the Killer Whale (p. 110) is distinctly black and white.

Short-finned Pilot Whale

REPRODUCTION: These whales mate successfully throughout the year. The

OCEAN DOLPHINS

Short-finned Pilot Whale

Globicephala macrorhynchus

DESCRIPTION: The Short-finned Pilot Whale is a small, black or dark gray whale. It has a bulbous head and an upward-slanting mouth, and some individuals have a light gray streak rising diagonally behind the eye. The dorsal fin is variable in shape, but it is usually arched strongly backward, with a rounded tip and a concave trailing edge. This whale has a light, W-shaped patch on its underside between its flippers, like the False Killer Whale (p. 104), and a light gray, oval patch on the underside forward of the tail stock. The flippers are pointed and strongly arched backward, and they are equal to 14 to 19 percent of the body length. The flukes are strongly pointed and notched in the middle. Short-finned Pilot Whales found in the Northern Hemisphere are typically larger than those in the Southern Hemisphere, and males are much larger than females.

BLOW: This whale makes a strong blow that is visible in calm weather.

OTHER DISPLAYS: The most commonly seen behavior of the Short-finned Pilot Whale is logging, in which entire pods loll at the surface and are approachable by boats. This pilot whale rarely breaches, but it may be seen "porpoising," lob-tailing and spy-hopping.

OCEAN DOLPHINS

TOOTHED WHALES

RANGE: found in open ocean in tropical and warm-temperate waters.

OTHER NAMES: Pothead Whale, Pacific Pilot Whale, Blackfish.

STATUS: common.

ADULT LENGTH: up to 23 ft; average 19 ft.

ADULT WEIGHT: up to 4 tons; average 2.5 tons.

BIRTH LENGTH: 4½–6 ft.

BIRTH WEIGHT: about 140 lb.

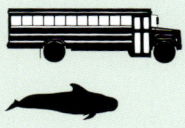

GROUP SIZE: The typical group size is 10 to 30 individuals, though certain circumstances may attract several hundred pilot whales to one place at one time.

REPRODUCTION: Length at sexual maturity varies greatly, but for females the range is 11–13 feet, and for males it is 14–18 feet. The gestation period for this species is not accurately known, but estimates range from 12 to 15 months. Populations in the Northern Hemisphere appear to calve in fall and winter. Females nurse their calves for nearly 2 years.

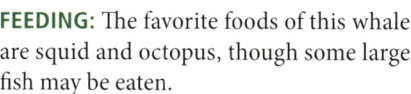

FEEDING: The favorite foods of this whale are squid and octopus, though some large fish may be eaten.

SIMILAR SPECIES: The Long-finned Pilot Whale (p. 108) has longer, more tapered flippers; the False Killer Whale (p. 104) has a less-bulbous melon and a more upright and slender dorsal fin; the Killer Whale (p. 110) is distinctly black and white.

Long-finned Pilot Whale

Long-finned Pilot Whale
Globicephala melas

DESCRIPTION: The Long-finned Pilot Whale is dark gray or black, with a signature anchor patch on the chest. The anchor patch is long and narrow and often continues to a white patch on the lower belly. Some individuals have a gray saddle patch behind the dorsal fin. Its long, pointed flippers and stumpy dorsal fin have a strongly swept-back appearance. The tail stock is thick, dwarfing the narrow tail flukes. Its melon is bulbous and often extends forward of the tip of the indistinct beak.

BLOW: This whale makes a strong but low, bushy blow that is sometimes visible.

OTHER DISPLAYS: Pilot whales engage in a variety of behaviors at the surface. They spy-hop, breach, log, lob-tail and "porpoise" through the water. Before a dive they show their tail flukes. This species is known for mass strandings, which usually result in numerous deaths. They are very social whales, and if one member of a group has stranded, the others appear to come to help but often end up stranding as well. The initial cause of the strandings is not known, but researchers theorize that the whales' echolocation is disrupted by shallow coastlines.

GROUP SIZE: Pods are tight and usually have 10 to 20 members, but these small pods are often part of a larger, loose group with over 100 individuals. Gatherings of several hundred pilot whales sometimes form, but for unknown reasons. These pilot whales

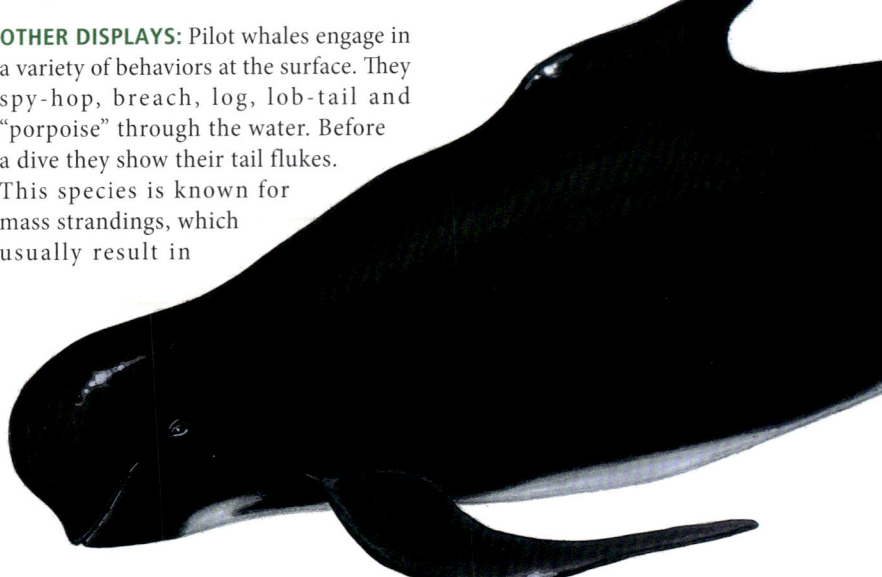

OCEAN DOLPHINS

RANGE: two distinct populations known; in the North Atlantic, ranges from Iceland to South Carolina.
OTHER NAMES: Pothead Whale, Blackfish.
STATUS: probably stable.
ADULT LENGTH: up to 25 ft; average 22 ft.
ADULT WEIGHT: up to 5000 lb; average 3800 lb.
BIRTH LENGTH: about 6 ft.
BIRTH WEIGHT: about 165 lb.

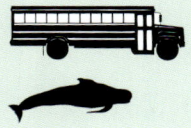

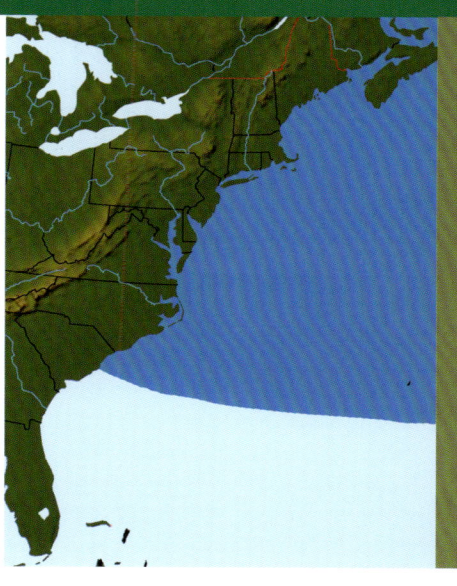

often mix with Bottlenose (p. 74) and Atlantic White-sided (p. 92) dolphins.

FEEDING: Long-finned Pilot Whales primarily feed on squid, but they will take medium-sized fish when available. They feed mainly at night, making dives that last about 18 minutes at depths of about 2600 feet.

REPRODUCTION: Mating and calving peaks in the summer months, and females give birth to 1 calf after a gestation of 12 to 16 months. Nursing lasts 1½ to 3½ years. Females become sexually mature when they are about 8 years old, and they may go as long as 6 years between calves. Males are sexually mature at 12 or 13 years old.

SIMILAR SPECIES: The Short-finned Pilot Whale (p. 106) has shorter flippers; the Pygmy Killer (p. 102), False Killer (p. 104) and Melon-headed (p. 100) whales are all much smaller and have a less-bulbous melon; the Killer Whale (p. 110) is distinctly black and white.

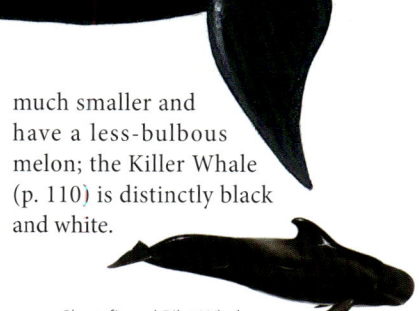

Short-finned Pilot Whale

OCEAN DOLPHINS

Killer Whale
Orcinus orca

The Killer Whale, also known as the Orca, with its striking colors and intelligent eyes, has fascinated humans for centuries. Long revered by the indigenous peoples of North America, the Killer Whale is now an icon for such concerns as biodiversity protection and contact with non-human intelligence. Images of this white-and-black giant can be found on everything from coffee cups to international conservation documents.

The Killer Whale lives in every ocean of the world, from the cold polar seas to the warm equatorial waters, and it is one of the most widely distributed mammals on Earth. Uncontested as a top predator in the oceans, the Killer Whale feeds on a wider variety of creatures than any other whale. It is regarded as an intelligent yet fearsome creature—it is the lion that rules the seas.

Studies indicate that there are three distinct types of Killer Whale: "transients," "residents" and "offshores." The three kinds are distinguishable by appearance and behavior, but the differences are subtle. Transient Killer Whales tend to be larger than residents, and they have taller, straighter dorsal fins. The transients live in smaller pods, from one to seven individuals, and they have large home ranges; resident Killer Whales have seasonally small ranges and travel along predictable routes. Transients are more likely to feed on other sea mammals, they dive for up to 15 minutes, they make erratic direction changes while traveling, and they do not vocalize as

OCEAN DOLPHINS

RANGE: found in all the world's oceans and most adjoining seas; the most widespread of all whales.
OTHER NAMES: Orca.
STATUS: vulnerable; locally stable.
ADULT LENGTH: up to 32 ft; average 28 ft.
ADULT WEIGHT: up to 11 tons; average 7.5 tons.
BIRTH LENGTH: 6–8 ft.
BIRTH WEIGHT: about 400 lb.

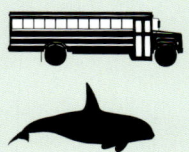

much as residents. Resident Killer Whales feed mainly on fish, are highly vocal and rarely dive for longer than three or four minutes. Offshore Killer Whales resemble the residents in appearance, but they usually live farther out at sea. As well, offshore Killer Whales tend to travel in groups of up to 100 individuals. Much more research is needed to accurately describe the offshore Killer Whales. At this time, only the transients have been found along the east coast. They are not very common here, but the best sites to encounter them include Cape Cod and Cape Ann.

Killer Whales have never been hunted heavily by humans. Some hunting has occurred in the past several decades, but it has not threatened the total population. One form of hunting that has taken many Killer Whales and their close cousins, the Bottlenose Dolphins (p. 74), is live hunting. Killer Whales are taken from the wild and introduced into the world aquarium trade, where they are expected to learn and perform tricks for the public. The capture of these animals for this purpose has caused much controversy, mainly because many people consider cetaceans to be extremely intelligent animals and think that keeping them in a closed and cramped aquarium is unjust. Much of what we have learned about cetacean intelligence and biology, however, has come from aquarium studies, and this knowledge helps us to better understand and protect whales in the wild.

OCEAN DOLPHINS

DESCRIPTION: The Killer Whale is unmistakable: its body is black, its undersides and lower jaw are white, there are white patches behind the eyes and on its sides, its large flippers are paddle-shaped and its dorsal fin is tall and triangular. The dorsal fin varies in shape and size from whale to whale, and it often bears scars and nicks. Old males may have a fin as tall as 6 feet, and the fin of some old individuals is wavy when seen straight on. Females have a smaller, more curved dorsal fin than males. Behind the dorsal fin there is often a uniquely shaped gray or purplish "saddle." Scientists and whale watchers rely on these varying dorsal fin and saddle characteristics to identify individual whales. A Killer Whale's eye is located below and in front of the white facial spot. The snout tapers to a rounded point. The flukes are dark on top and whitish below, and they have pointed tips, concave trailing edges and a distinct notch in the middle. Males are much larger than females.

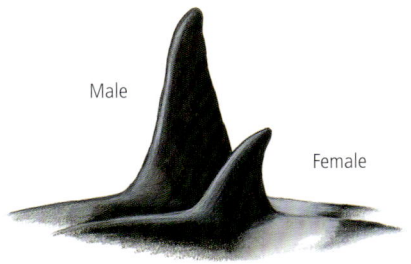
Male
Female

OCEAN DOLPHINS

FEEDING: Killer Whales feed on a wider variety of animals than any other whale, partly because of their global distribution. Several hundred animal species are potential prey to these top predators of the sea, including, but not limited to, seals, other whales and dolphins, dugongs, fish, sea turtles and birds. Even land mammals may be eaten—there are records of pods killing and eating moose and caribou that swim across narrow channels and rivers in Canada and Alaska.

REPRODUCTION: Mating probably occurs between individuals of a pod, but because "super-pods" occasionally form, paternity is almost impossible to ascertain without genetic analysis. Males reach maturity when they are about 19 feet long, and females when they are about 16 feet long. Winter appears to be the peak calving period, and gestation is believed to take 12 to 16 months.

SIMILAR SPECIES: The Killer Whale's black and white coloration is unmistakable. The False Killer Whale (p. 104) is much smaller; the Short-finned (p. 106) and Long-finned (p. 108) pilot whales have a bulbous head and longer, tapered flippers.

BLOW: In cool air, the Killer Whale makes a low, bushy blow.

OTHER DISPLAYS: Killer Whales are extremely acrobatic whales for their size. They are inquisitive and often approach boats, apparently to get a better look at the humans on board. They are often seen breaching clear out of the water, lob-tailing, flipper-slapping, logging and spy-hopping. They may speed-swim, or "porpoise," with their entire body leaving the water at each breath.

GROUP SIZE: Killer Whales usually travel in pods of 3 to 25 individuals. Some social gatherings may attract several pods at one time.

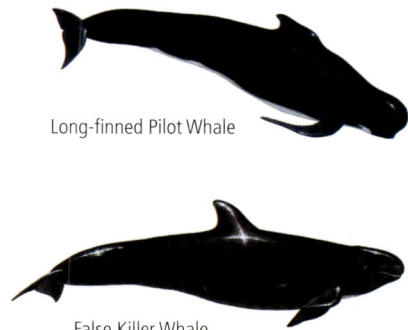

Long-finned Pilot Whale

False Killer Whale

TOOTHED WHALES

113

PORPOISES

Harbor Porpoise
Phocoena phocoena

The little Harbor Porpoise is the most widespread member of its family. It favors coastal waters, such as estuaries, shallow bays and tidal channels, and it has even been seen short distances up rivers. It is rarely found in waters deeper than 1000 feet and seems to prefer water no deeper than 650 feet. Along the east coast, sightings are common in water less than 200 feet deep.

In the past, this porpoise was a familiar sight to boaters and sailors of coastal waters, but its population is declining. In some regions, alarming numbers of Harbor Porpoises have washed up on shorelines either dead or dying. Although the reasons are not clear, many scientists suspect that high levels of toxins and pollutants impair the immune systems of these and other cetaceans, which makes them more susceptible to life-threatening diseases.

Drowning deaths are another main reason why Harbor Porpoises are declining. Because of this porpoise's feeding habits, it is frequently caught in bottom-set gill nets, a kind of net that sits deep in the water like a curtain attached to the bottom. Almost every kind of deep net can trap and drown a cetacean, but the most sinister nets are the ones that have been thrown away or lost by fishermen. An untended net catches numerous cetaceans, sea turtles, fish and other sea creatures until it is so heavy that it sinks to the bottom.

The solution to this problem is not easy, but leading

PORPOISES

RANGE: found in cold-temperate and subarctic waters of the Northern Hemisphere.

OTHER NAMES: Common Porpoise, Puffing Porpoise.

STATUS: declining.

ADULT LENGTH: up to 6½ ft; average 5 ft.

ADULT WEIGHT: up to 150 lb; average 130 lb.

BIRTH LENGTH: 26–35 in.

BIRTH WEIGHT: about 11 lb.

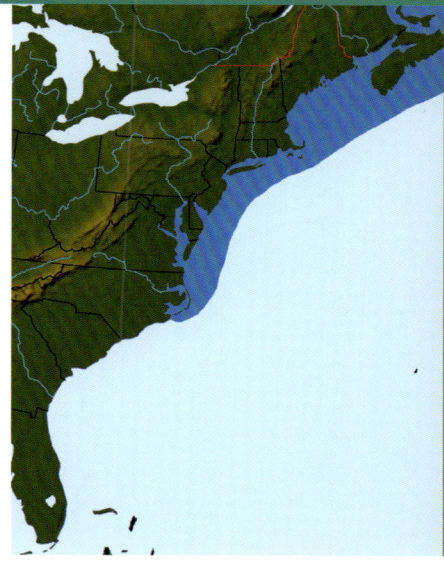

TOOTHED WHALES

researchers have been working on ways to warn animals, particularly porpoises, where the nets are. Some sounds repulse Harbor Porpoises, and now a device called a "pinger" can be specially made and fitted to the nets. The pinger emits a sound that the porpoises do not like, resulting in the porpoises steering clear of the net. In the future, devices of this kind may help save the lives of thousands of porpoises and other cetaceans.

DESCRIPTION: This small porpoise is the smallest cetacean in the region. It is mainly dark or slate gray over the back, fading to white underneath. In some individuals the color change is gradual, but in others the change is well defined, especially on the tail stock. Even on one individual, the color pattern may not be the same from one side to the other. The face is small, the head tapers gently and the mouth angles slightly upward. This porpoise has black "lips," and 1 or 2 black streaks extend back to its all-black flippers. The dorsal fin and flukes are black as well. The flukes are pointed

PORPOISES

slightly backward, and they are notched in the middle. Upon close examination, you can see little bumps, or tubercles, on the leading edges of the dorsal fin and flippers.

BLOW: Although Harbor Porpoises are difficult to observe because they rarely show much of themselves abovewater, they make a remarkable sound when they breathe. When a Harbor Porpoise breaks the surface, it makes a quick sneezing sound, which has earned it the nickname "Puffing Porpoise."

OTHER DISPLAYS: Like other porpoises, the Harbor Porpoise shows little of itself abovewater. It rarely performs acrobatics like dolphins do, and it is not as fast or active as a dolphin. Only its comical sneezing sound attracts the attention of whale watchers. Harbor Porpoises do not like intruders, however, and if they are

PORPOISES

GROUP SIZE: Harbor Porpoises usually live in small groups of 2 to 5 individuals. Some groups have up to 15 members, and good feeding waters can attract from 50 to several hundred porpoises.

FEEDING: Harbor Porpoises feed in midwater or on the bottom, and their main foods include small schooling fish, such as anchovies or herring. They rarely venture into water more than 650 feet deep.

REPRODUCTION: Both male and female Harbor Porpoises mature sexually between the ages of 3 and 4. Mating usually occurs in early summer, and calving is 11 months later. The newborns are dull brown in color, and for the first few hours they have birth lines or creases circumscribing their bodies. This species is not long-lived; most individuals die before they are 10 years old.

SIMILAR SPECIES: The larger Atlantic White-sided Dolphin (p. 92) has a distinct tricoloured pattern; the larger White-beaked Dolphin (p. 96) has more distinct markings and a larger dorsal fin.

uncomfortable and want to hide they will lie silently and motionlessly just below the surface of the water. This behavior also occurs at night, and it probably helps them remain undetected by predators. On rare occasions, Harbor Porpoises have been seen bow-riding. When they swim quickly, they may "porpoise" in and out of the water.

Atlantic White-sided Dolphin

White-beaked Dolphin

TOOTHED WHALES

BEAKED WHALES

Cuvier's Beaked Whale
Ziphius cavirostris

DESCRIPTION: This medium-sized whale has a highly variable color range that includes cream, beige, brown and purplish or reddish black. The color pattern is swirled and blotchy, and most individuals have spots and scars along their sides and undersides. The head is triangular, with an indistinct beak and 2 small teeth on the protruding lower jaw. The flippers are small and close to the head. The dorsal fin is small and close to the tail. The flukes are pointed, only slightly notched in the middle and broad—they measure almost as much as ¼ of the length of the body. Females are generally larger than males.

BLOW: The blow is usually invisible, but it may be seen after a long dive. It is low and bushy, and it generally points forward and to the left.

OTHER DISPLAYS: This beaked whale breaches infrequently, rising vertically and completely out of the water and falling ungracefully back in. It usually avoids boats and whale watchers, but there are a few notable cases where inquisitive individuals have approached boats.

GROUP SIZE: When solitary Cuvier's Beaked Whales are seen, they are usually old males. Normally, these whales form groups of 2 to 10 members, but pods as large as 25 have been recorded.

FEEDING: These whales make deep dives of 20 to 40 minutes to feed on deep-sea fish and squid. They are rarely seen close to land.

118

BEAKED WHALES

RANGE: found throughout the world, in all seas except cold polar waters; the most widespread member of its family and probably the most common.

OTHER NAMES: Goose-beaked Whale, Cuvier's Whale.

STATUS: insufficiently known.

ADULT LENGTH: up to 23 ft; average 20 ft.

ADULT WEIGHT: up to 3.5 tons; average 2.5 tons.

BIRTH LENGTH: 6½–9½ ft.

BIRTH WEIGHT: about 550 lb.

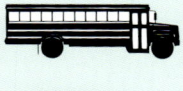

REPRODUCTION: Sexual maturity for both males and females is reached at about 75 percent of the adult length. The peak periods for mating and calving are unknown.

TOOTHED WHALES

SIMILAR SPECIES: The Sowerby's Beaked Whale (p. 122) is smaller, and its teeth erupt ⅔ back along the jawline; the Gervais's Beaked Whale (p. 126) has teeth that erupt ⅓ back along the jawline; the True's Beaked Whale (p. 128) is smaller.

Sowerby's Beaked Whale

BEAKED WHALES

North Atlantic Bottlenose Whale

Hyperoodon ampullatus

DESCRIPTION: These large beaked whales are usually gray, olive or brown in color. Their sides are lighter and the belly is pale gray or white. Sometimes the color is blotchy or swirly over the back. Both sexes have a pronounced melon, though it is somewhat smaller in the female. There are 2 small, conical teeth at the tip of the lower jaw. The strongly falcate dorsal fin is 12–16 inches tall and located ⅔ of the way down the length of the dorsum. Their tail flukes lack the notch in the center. Males are about 3 feet longer than females.

BLOW: When visible, the low, bushy blow is angled slightly forward.

OTHER DISPLAYS: This whale has been known to approach and circle slow-moving boats, sometimes even for hours. Lob-tailing and breaching are infrequent behaviors. Unlike most other whales, they do not commonly show their flukes before a deep dive.

GROUP SIZE: These whales are usually found in groups of 4 to 10 individuals, though it is not uncommon for 25 or 30 to be seen together.

FEEDING: North Atlantic Bottlenose Whales feed primarily on squid. Fish, cuttlefish, and bottom-dwelling invertebrates such as sea stars and sea cucumbers are also consumed. Like other beaked whales, they feed by sucking their prey into their mouths.

BEAKED WHALES

RANGE: inhabits most of the North Atlantic; is a deepwater species; best-known population is found in waters over the large submarine canyon known as the Gully, just east of Maine.

OTHER NAMES: Northern Bottlenose Whale, Bottlehead.

STATUS: unknown.

ADULT LENGTH: up to 33 ft; average 29 ft.

ADULT WEIGHT: up to 8 tons; average 7 tons.

BIRTH LENGTH: about 12 ft.

BIRTH WEIGHT: unknown.

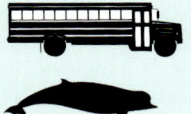

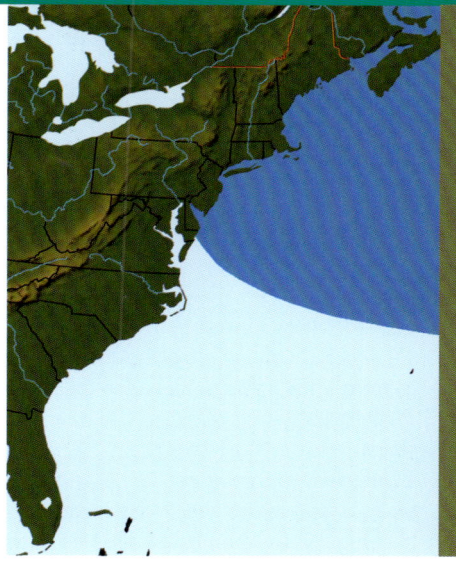

REPRODUCTION: Between April and June, a female gives birth to 1 calf after a 12-month gestation. The calf is weaned at 1 year old and reaches maturity at 7 to 9 years old. These whales are known to live for more than 35 years.

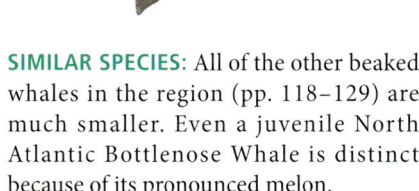

SIMILAR SPECIES: All of the other beaked whales in the region (pp. 118–129) are much smaller. Even a juvenile North Atlantic Bottlenose Whale is distinct because of its pronounced melon.

Sowerby's Beaked Whale

Mesoplodon bidens

DESCRIPTION: Sowerby's Beaked Whales are generally bluish gray or charcoal gray in color, with lighter undersides and belly. There are often scars from cookiecutter sharks along their sides, giving them a mottled appearance. Their snout is tapered, and males have a slightly arched lower jaw with 2 conical teeth, 1 on each side about ⅔ back along the jawline. The teeth of females do not erupt, and their jawline is straight. The melon is only a slight bulge on the forehead, and it is more pronounced in males. The falcate dorsal fin is located quite far back, and the tail flukes lack the central notch.

BLOW: The blow is small and inconspicuous, even in cool weather.

OTHER DISPLAYS: These whales are known to breach, leaving the water at steep angles. Sometimes they clear the water completely before splashing back in. Occasional strandings are reported.

GROUP SIZE: Sowerby's Beaked Whales form small groups of 3 to 10 individuals.

FEEDING: This species feeds by suction and preys mainly on squid and some fish. It favors deep waters past the shelf, but it is known to visit and feed in waters over the deep submarine canyon known as the Gully off the coast of Nova Scotia, east of Maine. Some stranded, dead individuals were found to have swallowed plastic items, which likely caused or contributed to their death. Plastics, especially plastic bags, resemble the squid on which they feed. Other beaked whales are similarly affected.

BEAKED WHALES

RANGE: inhabits waters beyond the shelf edge in the North Atlantic from Massachusetts to Labrador and north to Norway.

OTHER NAMES: North Atlantic Beaked Whale, North Sea Beaked Whale.

STATUS: unknown.

ADULT LENGTH: up to 18 ft; average 17 ft.

ADULT WEIGHT: up to 2900 lb; average 2600 lb.

BIRTH LENGTH: 8–9 ft.

BIRTH WEIGHT: about 400 lb.

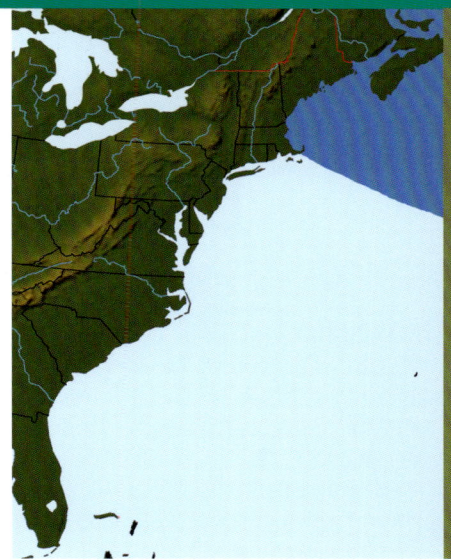

REPRODUCTION: Little is known of the reproduction of these whales, but females are estimated to reach maturity at about 7 years old. Females give birth to 1 calf probably every 2 to 3 years.

SIMILAR SPECIES: The Cuvier's Beaked Whale (p. 118) is larger and has teeth that erupt near the tip of the beak; the Gervais's Beaked Whale (p. 126) has teeth that erupt ⅓ back along the jawline; the True's Beaked Whale (p. 128) has teeth that erupt near the tip of the beak.

TOOTHED WHALES

Cuvier's Beaked Whale

True's Beaked Whale

Blainville's Beaked Whale

Mesoplodon densirostris

DESCRIPTION: This species is slate gray or bluish over the back, and it has pale undersides. The body may have extensive blotches and scars, giving it a mottled appearance. Perhaps the most distinguishing feature of this whale is the male's 2 teeth, which are quite large. They erupt from the lower jaw, which arches upward around the upper jaw, and, because the forehead is flattened, the tips of the teeth are often higher than the top surface of the snout. The female also has an arched lower jaw, but her teeth do not erupt and are not visible. The flippers are small, the dorsal fin is triangular and somewhat falcate, and the tail flukes are slender and swept backward. The peculiar Latin name (and an alternate common name) of this species refers to its dense jawbones. When this whale was first described, a sample of bone from the jaw was found to be of very high density. The species has since been discovered to have the densest bones in the animal kingdom.

BLOW: This whale makes a small, inconspicuous blow that is directed forward. The best chance to see its blow is on a calm, cool day.

OTHER DISPLAYS: These whales are not acrobatic, and the extent of their surface behavior is breathing and resting. When breathing before a deep dive, a Blainville's Beaked Whale will surface several times during a period of 20 to 45 minutes. The last time it surfaces before a deep dive, the forward roll of its dive sequence will be strongly arched.

BEAKED WHALES

RANGE: found in warm-temperate waters around the world; distribution probably not continuous; population in the North Atlantic not well documented.

OTHER NAMES: Dense Beaked Whale, Tropical Beaked Whale, Atlantic Beaked Whale.

STATUS: insufficiently known.

ADULT LENGTH: up to 20 ft; average 17 ft.

ADULT WEIGHT: up to 1.2 tons; average 1 ton.

BIRTH LENGTH: 6–8½ ft.

BIRTH WEIGHT: about 130 lb.

TOOTHED WHALES

GROUP SIZE: These beaked whales usually live singly or in small groups of up to 6 members. Sometimes groups as large as 12 are seen.

FEEDING: Little is known about this whale's habits, but it probably feeds on squid and deepwater fish.

REPRODUCTION: Average age of sexual maturity is thought to be 8 or 9 years, but little else is known about this whale's reproductive habits.

SIMILAR SPECIES: The other beaked whales in the region (pp. 118–129) lack the strongly arched lower jaw.

Gervais's Beaked Whale
Mesoplodon europaeus

DESCRIPTION: The Gervais's Beaked Whale is primarily charcoal gray with lighter undersides. There may be many scars on the back and cookiecutter shark bites over the sides. The small, tapered head has a small, slightly bulging melon. The jawline is very straight, and males have 2 small teeth that erupt in the lower jaw about ⅓ of the way back along the jawline from the tip of the beak. The falcate dorsal fin is wide at the base and located quite far back along the dorsal ridge. Of the individuals studied, females are slightly larger than males.

BLOW: The small, bushy blow is rarely visible.

OTHER DISPLAYS: These whales tend to be very skittish and are rarely observed. Known at first from stranded individuals, this species was seen alive for the first time in 1998. Since then there have been only a handful of confirmed live sightings. Stranded individuals have defined the range for the species.

GROUP SIZE: Like other beaked whales, Gervais's Beaked Whales may be found individually or in small groups of 3 to 10 members.

FEEDING: Gervais's Beaked Whales feed primarily on deepwater squid. Fish and some invertebrates are also consumed, but to a lesser extent. This species, like other beaked whales, is known to consume plastic debris that resembles squid in the water. Plastic debris is the cause of death in many of the stranded individuals that have been studied.

BEAKED WHALES

RANGE: inhabits tropical and temperate waters of the Atlantic more or less from Cape Cod to Brazil.

OTHER NAMES: Gulf Stream Beaked Whale, European Beaked Whale, Antillean Beaked Whale.

STATUS: unknown.

ADULT LENGTH: up to 18 ft; average 15 ft.

ADULT WEIGHT: about 2600 lb.

BIRTH LENGTH: 5–7 ft.

BIRTH WEIGHT: about 175 lb.

TOOTHED WHALES

SIMILAR SPECIES: The Sowerby's Beaked Whale (p. 122) has a longer beak and teeth that erupt ⅔ back along the jawline; the Cuvier's Beaked Whale (p. 118) is larger and has teeth that erupt near the tip of the beak; the True's Beaked Whale (p. 128) has a straighter beak and teeth that erupt near the tip of the beak.

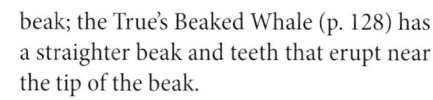

Sowerby's Beaked Whale

REPRODUCTION: Little is known about the reproduction of this species. Like other beaked whales, the female will probably give birth to 1 calf every 2 to 3 years. The lifespan is at least 25 years, with upper estimates approaching 50 years.

True's Beaked Whale
Mesoplodon mirus

DESCRIPTION: The True's Beaked Whale is mainly dark gray or brown, with lighter sides and a pale, nearly white belly. Some individuals may have light patches around and behind the eye, and sometimes on the tail stock as well. Cookiecutter shark bite scars are common on older individuals, and long, scratchy scars may be seen on the backs of males. These are probably inflicted by other males during dominance fights in the mating season. The beak is relatively short, and the jawline is only slightly curved. A mature male has 2 teeth that erupt close to the tip of the beak. The melon bulges slightly between the beak and the blowhole. Its Latin name *mirus* means "wonderful" and was given to the whale by Frederick True, who was excited by his discovery of this new species. Further studies may reveal that the two apparently separate populations of True's Beaked Whales are actually distinct species.

BLOW: The blow is small and inconspicuous, even in cool weather.

OTHER DISPLAYS: When surfacing, True's Beaked Whales may show their entire beak. These whales have a low profile and have rarely been seen alive, and breaching is the only display that has been observed. Anecdotal evidence suggests that an injured individual is tended by at least one of its group members, a trait also described in other beaked whales.

GROUP SIZE: The most common group size appears to be 3 members, but single individuals have been encountered. Probably groups of more than 3 occur as well.

FEEDING: Like other members of its genus, the True's Beaked Whale feeds on

BEAKED WHALES

TOOTHED WHALES

RANGE: two populations: one in the Atlantic from Nova Scotia to the Bahamas; the other in the Indian Ocean between South Africa and Australia.

OTHER NAMES: Wonderful Beaked Whale.

STATUS: unknown.

ADULT LENGTH: up to 17½ ft; average 16 ft.

ADULT WEIGHT: up to 3000 lb; average 2500 lb.

BIRTH LENGTH: 6½–8 ft.

BIRTH WEIGHT: about 300 lb.

deepwater squid and some fish. Although this is a deepwater species that feeds on deepwater prey, individuals have been sighted in shelf waters.

REPRODUCTION: Like other beaked whales in this family, females probably give birth to 1 calf every 2 to 3 years. Nothing is known of gestation and nursing periods, and the details of mating and birthing remain a mystery.

SIMILAR SPECIES: The Sowerby's Beaked Whale (p. 122) has teeth that erupt ⅔ back along the jawline; the Cuvier's Beaked Whale (p. 118) is larger; the Gervais's Beaked Whale (p. 126) has a more down-turned beak and has teeth that erupt ⅓ back along the jawline.

Sowerby's Beaked Whale

Pygmy Sperm Whale
Kogia breviceps

DESCRIPTION: The overall body color is slate blue, dark gray or even brownish, and the undersides are nearly white. Like its close relative, the Dwarf Sperm Whale (p. 132), this small whale is almost shark-like in appearance. It has a protruding snout and an underslung lower jaw, and many individuals have a streak behind the eye that resembles a gill slit. The blowhole is on top of the head and is displaced slightly to the left. The dorsal fin is tiny, slender and curved, and the flippers are broad and almond-shaped. The tail flukes are convex and notched, with distinctly pointed tips.

BLOW: These whales make small, inconspicuous blows that are visible only on clear, calm days and at close range.

OTHER DISPLAYS: When they surface to breathe, Pygmy Sperm Whales are very inconspicuous, and they usually drop out of sight immediately. These whales may occasionally breach, and they typically fall back into the water with little grace. They sometimes release a cloud of dark intestinal fluid that probably serves to either distract intruders or confuse prey, or both.

GROUP SIZE: Pygmy Sperm Whales live in small groups of 3 to 6 members. On rare occasions they have been observed singly or in groups as large as 10 individuals.

FEEDING: Pygmy Sperm Whales feed mainly in deep water, catching such creatures as squid, fish and some crustaceans.

DWARF SPERM WHALES

RANGE: widespread in deep tropical, subtropical and temperate waters; distribution is patchy and insufficiently known.

OTHER NAMES: Lesser Sperm Whale, Lesser Cachalot, Short-headed Whale.

STATUS: insufficiently known.

ADULT LENGTH: up to 12 ft; average 10 ft.

ADULT WEIGHT: up to 900 lb; average 780 lb.

BIRTH LENGTH: about 4 ft.

BIRTH WEIGHT: about 120 lb.

TOOTHED WHALES

REPRODUCTION: Length at sexual maturity is believed to be about 9 feet for Pygmy Sperm Whales. Calving and mating may occur year-round, but there is little accurate data for this species.

SIMILAR SPECIES: The Dwarf Sperm Whale (p. 132) has a larger dorsal fin and a more pointed snout; the Risso's Dolphin (p. 98) has distinct scarring over its head and body.

Risso's Dolphin

Dwarf Sperm Whale

Dwarf Sperm Whale
Kogia sima

DESCRIPTION: This unusual whale resembles a slightly fat shark. Its snout is short and squarish, and its mouth is tiny and underslung. There is a white streak behind the eye that resembles a gill slit on a shark. Overall, the body is dark or bluish gray, with lighter undersides. The flippers are short and broad, and the dorsal fin is pointed and has a concave trailing edge. The flukes are broad and pointed, and they are notched in the middle. The blowhole is set above the eye and is slightly displaced to the left.

BLOW: The Dwarf Sperm Whale does not make a distinct blow.

OTHER DISPLAYS: When this whale rises to breathe, it does not roll forward into a dive like other whales; instead, it takes a quick breath and then simply drops out of sight. It may be seen resting at the surface, and it occasionally breaches—rising vertically out of the water and splashing back in haphazardly. When startled, it may release a cloud of dark intestinal fluid that presumably distracts an intruder while it dives. These clouds are probably also released to confuse prey.

GROUP SIZE: Dwarf Sperm Whales travel alone or in pairs. On occasion, they have been seen in groups of up to 10 individuals.

FEEDING: This whale has a small mouth and presumably eats small creatures. Its primary foods are thought to be cuttlefish, squid and other fish.

REPRODUCTION: The gestation period for this whale is about 9 months, and calving is at

DWARF SPERM WHALES

RANGE: seen in a number of locations throughout deep temperate, sub-tropical and tropical waters in both the Northern and Southern hemispheres; distribution probably patchy rather than continuous.

STATUS: insufficiently known; probably stable.

ADULT LENGTH: up to 9 ft; average 8 ft.

ADULT WEIGHT: up to 610 lb; average 450 lb.

BIRTH LENGTH: about 39 in.

BIRTH WEIGHT: 90–110 lb.

TOOTHED WHALES

its peak during the summer months. Both sexes reach maturity when they are 75 to 80 percent of their adult length.

SIMILAR SPECIES: The Pygmy Sperm Whale (p. 130) has a smaller dorsal fin and a blunt snout; the Risso's Dolphin (p. 98) has distinct scarring over its head and body.

Pygmy Sperm Whale

Risso's Dolphin

Sperm Whale

Physeter macrocephalus

The famous Sperm Whale is a definite record-breaker in the world of mammals. It is easily the deepest-diving mammal on Earth, and it can hold its breath for more than two hours. Sperm Whales seem not to have a maximum diving depth, because greater and greater dives keep being discovered. The restricting factor appears to be the time without oxygen, rather than the enormous pressure on the body. At 10,500 feet, which is believed to be the deepest recorded dive, the pressure on the body is a staggering 317 atmospheres, yet Sperm Whales are still able to function.

The Sperm Whale's biggest adaptation to surviving these pressures is that before it dives, it removes the air from its lungs to help minimize the difference in pressure between its body cavities and the surrounding water. Because the whale's body, like ours, is mainly water, it resists compression.

The mammoth head of the Sperm Whale is filled with spermaceti, an oil that may function to regulate the whale's buoyancy as it dives and surfaces. The other proposed theory is that the spermaceti serves to magnify or focus the whale's sound emission during echolocation. Perhaps the spermaceti performs both of these functions for the whale. Females have dramatically smaller heads than males, but the reason for this difference is not clearly understood. Females also dive deeply and rely on echolocation, so perhaps the

DIVE SEQUENCE

SPERM WHALES

RANGE: found in deep waters worldwide; although widespread, distribution is probably patchy.

OTHER NAMES: Cachalot, Anvil-headed Whale.

STATUS: endangered (USFWS); locally stable in some waters.

ADULT LENGTH: up to 69 ft; average 48 ft.

ADULT WEIGHT: up to 58 tons; average 35 tons.

BIRTH LENGTH: about 13 ft.

BIRTH WEIGHT: about 1 ton.

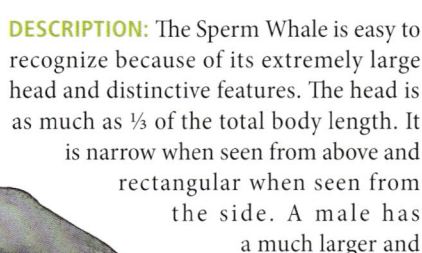

DESCRIPTION: The Sperm Whale is easy to recognize because of its extremely large head and distinctive features. The head is as much as ⅓ of the total body length. It is narrow when seen from above and rectangular when seen from the side. A male has a much larger and more rectangular head than a female. The eyes are relatively small and inconspicuous. The mouth and tip of the snout may be white or light gray, and scarring may be obvious. The slightly raised blowhole is at the end of the snout, and it is displaced to the left. The overall larger head of the male serves another purpose.

Sperm Whales are present all year in deep waters off the east coast, though they are sometimes found in shallower water in this region, too. Despite their presence, however, they are not reliably seen. Some tour operators may have current information on their whereabouts.

TOOTHED WHALES

SPERM WHALES

body color is brown, gray or nearly black, though the undersides are often lighter. The stubby flippers and broad, triangular tail flukes are dark on both sides, but scarring may lighten the edges. The trailing edge of the flukes is sometimes scalloped, and such variations in the tail are used by researchers to identify individuals. Instead of a true dorsal fin, the Sperm Whale has a distinct hump or knuckle where the fin would be. Several smaller knuckles continue down the dorsal ridge to the tail. The skin appears "pruney" from behind the eye to the tail stock. Males are much larger than females.

BLOW: Because the blowhole of the Sperm Whale is at the tip of the snout and displaced to the left, the blow sprays forward

Diving

SPERM WHALES

and to the left. No other whale makes a blow like this, so even from afar a Sperm Whale can be readily identified.

OTHER DISPLAYS: Sperm Whales, especially juveniles, may breach and lob-tail. When they return from a deep dive, they breathe at the surface 30 to 40 times before diving again. The first breath is very loud and explosive, but the subsequent breaths are quieter. Between dives, the whales usually stay at the surface for about 15 minutes, but sometimes they will remain near the surface for as long as an hour.

GROUP SIZE: Sperm Whales usually live in groups of 2 to 25 animals. Nursery colonies of females and young number 12 to 30 individuals, and bachelor colonies of many young males are also common. At times, assemblages of hundreds or even thousands of Sperm Whales have been reported.

FEEDING: The ability of the Sperm Whale to feed at great depths is unsurpassed by any other mammal. One of the deepest recorded dives is from two Sperm Whales that were believed to have reached 10,500 feet, where they had consumed bottom-dwelling sharks. The main foods eaten by Sperm Whales are squid and giant squid, octopus, large fish and sharks. Sperm Whales locate their food in the darkness of deep water by using their extremely sophisticated echolocation.

REPRODUCTION: Although males are sexually mature when they are about 10 years old, they are unlikely to mate for several more years. Females are sexually mature when they are 8 to 12 years old. Mating usually occurs in late winter or early spring, and single calves are born in summer or fall, following a gestation period of 14 to 15 months. The young may nurse for several years, even after they start eating solid food.

SIMILAR SPECIES: No other whale has a head as large as that of a Sperm Whale. The Humpback Whale (p. 62) has a similar dive sequence, but its knobby head is unmistakable.

Humpback Whale

TOOTHED WHALES

Seals, Otters and Manatees

Along the east coast, you may see several mammals other than whales, such as seals, otters and manatees. Although otters and seals are adapted to life in the ocean, they are unlike whales and manatees in that they are able to come ashore at any time. All seals, sea lions and otters are members of the order Carnivora, as are dogs, cats, bears and weasels, among others. The manatee is in its own order, Sirenia.

Hair Seal Family (Phocidae)

The hair seals are also referred to as either the "earless" seals or the "true" seals. As the former name suggests, they have no visible external ears. The best way to distinguish these seals is by body shape. The hindflippers of hair seals are permanently pointed backward—they are unable to rotate their hindlimbs forward to help support their body weight when they are on land. As a result, when hair seals are on land they move with clumsy undulations of their bodies. Once in the water, hair seals are graceful, lithe swimmers capable of high speeds and rapid turns. Their forelimbs are small compared to those of eared seals, but the flippers are strong enough that the seal can hold itself upright and tread water at the surface. Members of this family all have hair on their flippers and claws on all five digits of each flipper.

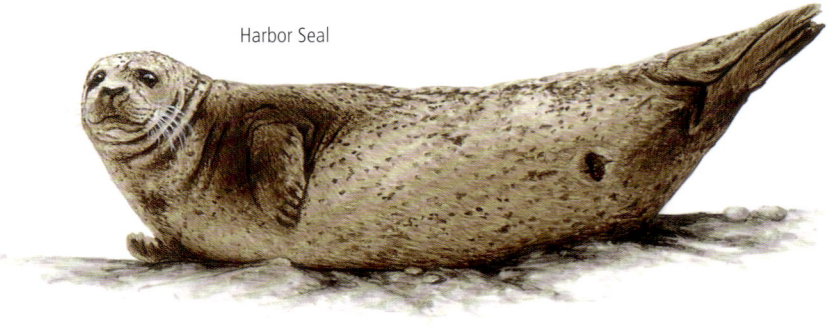

Harbor Seal

SEALS, OTTERS AND MANATEES

Otter Subfamily (Lutrinae)

Otters are a subgroup of the weasel family (Mustelidae), which also includes terrestrial carnivores such as minks, weasels, badgers and martens. There are two otters in North America: the Sea Otter and the Northern River Otter. The Sea Otter, which only lives on the west coast, was once heavily hunted for its thick fur. The Northern River Otter is similar to the Sea Otter, except that it is smaller and slimmer and has a longer tail. Although the Northern River Otter typically inhabits fresh water and is not strictly a marine mammal, it can be seen in some coastal waters.

Northern River Otter

Manatee or Sea Cow Family (Trichechidae)

Collectively called "sirens" in reference to their order, Sirenia, the representatives of this family are found in equatorial waters of the world. They inhabit marine wetlands, coastal marine waters, rivers, estuaries and swamps. Worldwide there are now only four species: three manatees and one dugong. The Steller's Sea Cow went extinct in the late 18th century, only 27 years after its discovery by Europeans, due to over-hunting. Several other species are known only from the fossil record. On the whole, sirens appear fat, but they are strong and extremely agile creatures. Their flippers are used for steering, while their paddle-like tail is used for propulsion. They make a variety of clicks and chirps that are audible to humans, and it is believed that this feature, coupled with their unusual tail shape, is what gave rise to the legend of mermaids.

West Indian Manatee

HAIR SEALS

Harbor Seal

Phoca vitulina

The inquisitive Harbor Seal is an increasingly common winter resident of the northern half of our east coast, with individuals seen as far south as Florida. This seal bespeckles rocky coastlines at almost any time of the day from September to May. It basks onshore either alone or in groups numbering up to the thousands.

Although many Harbor Seals may bask on rocks together, they pay very little attention to their neighbors and seldom interact. During the pupping season, females with newborn pups may congregate in a "nursery" in shallow water where the pups can sleep. The pups and the females sleep underwater, rising occasionally for a breath. These nursery groups are not socially interactive—they are simply a precautionary measure against predators. While most of the females and pups are sleeping, some are likely to be awake and watchful for danger. The same is true for hauled-out seals. Where several seals are together, the chances are good that there is always at least one individual awake and wary of approaching danger.

Harbor Seals tend to be cautious of humans, and if you approach them on land they are likely to dive immediately into the water. Many kayakers and boaters, however, have enjoyed watching inquisitive individuals that approach their boats for a better look. This kind of encounter is controlled by the seal: if it wants to see you, it will come closer, and if it is afraid of you, it will leave. Approaching a seal that has attempted to flee can cause unnecessary stress to the animal.

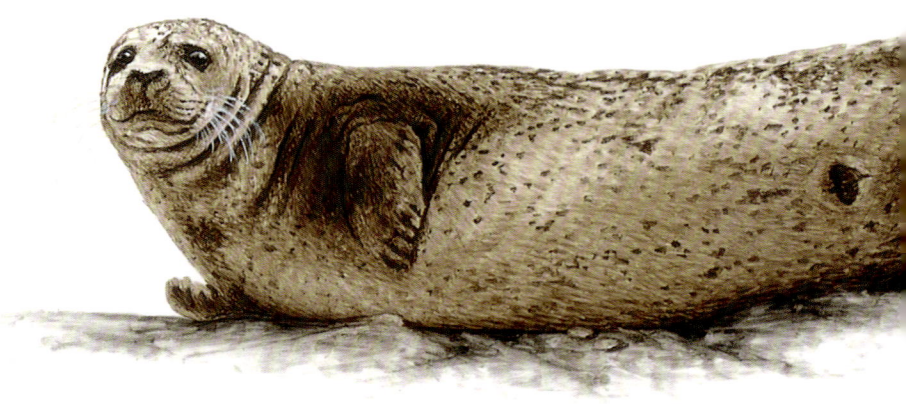

HAIR SEALS

RANGE: found along the northern coasts of North America, Europe and Asia; inhabits the U.S. east coast from Maine to Maryland; individuals spotted rarely in Florida, and only occasionally in Chesapeake Bay.
OTHER NAMES: Common Seal.
STATUS: common.
ADULT LENGTH: 4–6 ft.
ADULT WEIGHT: 110–310 lb.

Tracks

When the tide is out at night, Harbor Seals sleep high and dry at favorite haul-out sites. They frequently rest with their head and rear flippers lifted above the rock they are sleeping on. During the day, they can sleep underwater in shallow coastal water by resting vertically just above the bottom. Young pups commonly rest in this manner. Harbor Seals can go without breathing for nearly 30 minutes, and although they sometimes wake up to breathe, they frequently rise to the surface, take a breath and then sink back to the bottom without waking.

DESCRIPTION: The Harbor Seal is typically dark gray or brownish gray, with light, blotchy spots or rings, but some individuals may have the reverse color pattern— light gray or nearly white with dark spots. The undersides are generally lighter than the back. The outer coat is composed of stiff guard hairs about ½ inch long, which is the characteristic that gives seals of this

HAIR SEALS

family the name "hair seals." The guard hairs cover an undercoat of sparse, curly hair, about ¼ inch long, that helps provide insulation for the seal. Pups bear a spotted silvery or gray-brown coat at birth. The head is large, round and very dog-like, except that there are no visible ears. Each of the short foreflippers has long, narrow claws. Males are generally larger than females.

HABITAT: This near-shore seal is frequently found in bays and estuaries. Favored haul-out sites, which are traditionally used by many generations of Harbor Seals, commonly include intertidal sandbars, rocks and rocky shores. The Harbor Seal sometimes follows fish several hundred miles up major rivers, and there are even populations in some inland lakes—usually within 100 miles of the coast.

HAIR SEALS

FEEDING: Harbor Seals feed primarily on fish, such as rockfish, cod, herring, flounder and salmon. To a lesser extent, they also feed on mollusks (such as clams, squid and octopus) and crustaceans (such as crabs, shrimp and crayfish). Newly weaned pups seem to consume more shrimp and mollusks than do adults. Adult Harbor Seals have been seen taking fish from nets, and some have even entered fish traps to feed and then easily departed.

REPRODUCTION: Pupping colonies are usually located north of New England and on isolated islands. Gestation lasts 10 months, and a female will bear a single pup between April and July. The pups are weaned when they are 4 to 6 weeks old, after they have tripled their birth weight feeding on their mother's milk, which is more than 50 percent fat. Within a few days of weaning her pup, a female mates again. Harbor Seals become sexually mature anywhere from 3 to 7 years of age. Captive seals have lived for more than 35 years, but the typical lifespans are 20 years for males and 30 years for females.

SIMILAR SPECIES: The Gray Seal (p. 144) is larger; the Harp Seal (p. 146) has a distinct harp pattern over its back.

Gray Seal

Harp Seal

HAIR SEALS

Gray Seal
Halichoerus grypus

DESCRIPTION: These medium-sized seals have a very straight profile and nostrils that are quite far apart. They are mainly silvery gray or brown, with darker spots and blotches. Older males may have heavy scarring around their necks. Males are as much as 30 percent larger than females. Gray Seals are extremely vocal, and, in addition to bleats and gurgling sounds, they sing eerie songs. Their combined voices sound like a group of school children pretending to be ghosts.

HABITAT: Gray Seals inhabit the North Atlantic, specifically coastal waters where they can hunt for benthic fish at depths of 230–330 feet. There are haul-out sites and rookeries on both Atlantic shores. The largest breeding colony in the world is on Sable Island, just off the coast of Nova Scotia, and this protected colony is likely responsible for the increase of Gray Seals in New England.

FEEDING: Gray Seals feed on a wide variety of fish, especially benthic (bottom-dwelling) ones. Eels, cod, flatfish, herring and skates are commonly eaten, as well as octopus and lobster. These seals fast during the pupping and breeding season.

REPRODUCTION: Females come to shore in September, and pups are born from late September through November. Within a few weeks of giving birth, the female may mate again. Gestation is a little over 11 months, including a period of delayed implantation. Pups are born with white, fuzzy lanugo. They weigh 30 pounds at

HAIR SEALS

RANGE: found in coastal waters from Atlantic Canada to Maryland, as well as Iceland, Great Britain, Scandinavia and northern Europe.

OTHER NAMES: Atlantic Seal, Horsehead Seal.

STATUS: stable, with some local increase.

ADULT LENGTH: males up to 6½–9 ft; females up to 5¼–7¼ ft.

ADULT WEIGHT: males up to 530–700 lb; females up to 330–570 lb.

birth but grow very quickly because their mother's milk is about 60 percent fat. At about 1 month old, the pups shed their lanugo and develop their waterproof adult fur. Soon after, they head out to sea and feed for themselves. Gray Seals become sexually mature between 3 and 5 years old, and they can live for a long time; females up to 40 years and males up to 30 years.

SIMILAR SPECIES: The Harbor Seal (p. 140) is smaller, lacks the straight profile and has nostrils set closer together; the Harp Seal (p. 146) is smaller and has a distinct harp pattern over its back.

Harbor Seal

145

HAIR SEALS

Harp Seal
Phoca groenlandica

DESCRIPTION: Silvery white or grayish overall, this seal gets its name from the dark wishbone- or harp-shaped patch of fur over its back. Some individuals never fully develop their harp pattern and have just spots or blotches. These individuals are called "Spotted Harps." The pups are covered in fuzzy white fur for their first 12 days.

HABITAT: During winter months, Harp Seals haul out on ice in arctic and subarctic waters, where they give birth and later mate. The rest of the year is spent at sea; the peak of summer finds them in high arctic seas near Ellesmere Island. When feeding, they prefer waters shallower than 1000 feet deep.

FEEDING: Harp Seals regularly dive to depths of 500–650 feet to find their prey. They consume many kinds of small fish, such as capelin, cod, herring, sculpin, halibut, redfish and plaice. In fact, they are known to eat more than 65 different kinds of fish and at least 70 kinds of invertebrates (such as crab, lobster and shrimp).

REPRODUCTION: Pups are born in rookeries on ice in late February or March. They are 2½–3 feet long and have a fuzzy, yellow coat. After 2 or 3 days, the yellow coat is shed and the pup is covered in its famous white lanugo. The lanugo only lasts for 12 days, the period of the pup's most intense growth. The pups weigh about 22 pounds when they are born, but they are

HAIR SEALS

RANGE: circumpolar; migrates between arctic and subarctic waters; in North America, winter range is around Atlantic Canada to as far south as New England; summer range can reach as far north as Ellesmere Island and west into northern Hudson Bay.

OTHER NAMES: Greenland Seal, Saddle-backed Seal, Jumping Seal.

STATUS: common

ADULT LENGTH: 5¼–6¼ ft.

ADULT WEIGHT: 190–400 lb.

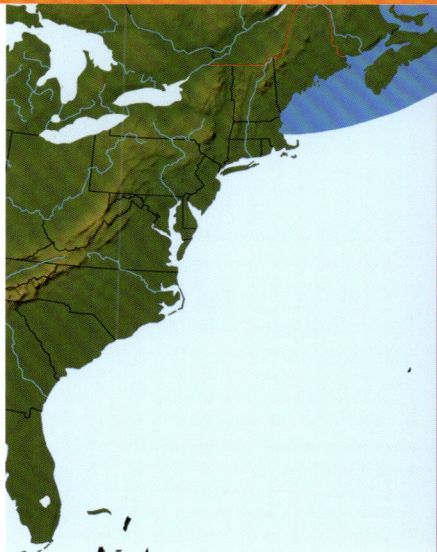

capable of putting on an astonishing 55 pounds in their first 10 to 12 days. They require this fat to survive; they are weaned at day 12, but they cannot swim until they are at least 4 weeks old. During these initial weeks, their newly stored fat keeps them alive. After the white lanugo is shed, the pup has a silvery coat with spots for up to 4 years. Sexual maturity is sometime after 4 years, at which time the adult color pattern is achieved. Females mate soon after giving birth, and males stay around the pupping colony for weeks to improve their chances of finding receptive females. Females have a delayed implantation of about 4 months, and their gestation is 7½ months, which results in pups being born at the same time every year.

SIMILAR SPECIES: The Gray Seal (p. 144) and Harbor Seal (p. 140) are usually darker and lack the harp pattern.

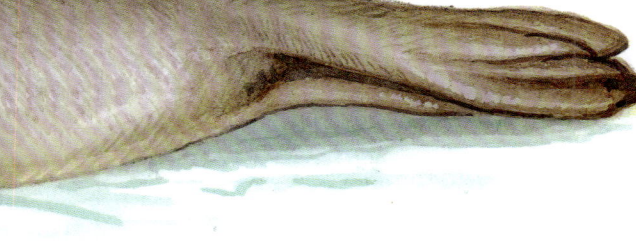

Northern River Otter
Lontra canadensis

Contrary to its name, the Northern River Otter is a year-round resident of coastal areas from Maine to Florida, in addition to its regular inland habitats. Although there are no true Sea Otters (*Enhydra lutris*) on the east coast, there was once the Sea Mink (*Neovison macrodon*) that lived in New England and the Maritimes. The Sea Mink was prized for its thick coat and was hunted to extinction by 1890. The Northern River Otter was also hunted, but its much larger range and population buffered it from extinction. Today, the Northern River Otter lives throughout the coast, now largely free of the hunting pressures that decimated the Sea Mink population. The competitive fur trade played an important role in North America's early economy, but it resulted in the endangerment, extirpation and extinction of many species.

OTTERS

RANGE: range extends from Labrador to Texas on the east coast, and from the Bering Strait to northern California on the west coast.

STATUS: uncommon.

ADULT LENGTH: 3½–4½ ft.

TAIL LENGTH: 12–20 in.

ADULT WEIGHT: 10–30 lb.

SEALS, OTTERS AND MANATEES

The Northern River Otter has a well-deserved reputation for playfulness. Otters often amuse themselves by rolling, sliding, diving or "body surfing," and they may also push and balance floating sticks with their noses or drop and retrieve pebbles for minutes at a time. They seem particularly interested in playing on slippery surfaces—they leap onto wet grass or mud with their forelegs folded close to their bodies for a streamlined toboggan ride. Unlike most members of the weasel family, river otters are social animals, and they will frolic together in the water and take turns sliding down slippery rocks.

With their streamlined bodies, rudder-like tails, webbed toes and valved ears and nostrils, river otters are well adapted for life in the water. When they emerge from water to clamber over rocks, there is a distinct serpentine appearance to their movement. In the water they are lithe and efficient at catching prey. Although otters generally cruise along slowly in the water

Back footprint   Front footprint

Tracks

149

OTTERS

by paddling with all four feet, they can dart after prey with the ease of a seal whenever hunger strikes. When an otter swims quickly, it propels itself mainly with vertical undulations of its body, hindlegs and tail. Otters can hold their breath for as long as five minutes.

Because of all its activity, the Northern River Otter leaves many signs of its presence when it occupies an area. Slides are the most obvious indicator of a resident otter. Also, despite its penchant for life in the water, an otter always defecates on land. Its scat is simple to identify—it is always full of fish bones and scales.

A river otter may make extensive journeys across land, and although this animal looks clumsy on land, it can easily outrun a human, even with its humped, loping gait. On slippery surfaces, such as wet grass or mud, the otter glides along, usually on its belly with its legs tucked either back or forward to help steer and push. On flat ground, slides are sometimes pitted with blurred footprints where the otter has given itself a push for momentum.

DESCRIPTION: This large, weasel-like carnivore has a dark brown back that looks black when it is wet. It is paler below, and the throat is often silver-gray. The head is broad and flattened, and it has small eyes, small ears and prominent, whitish whiskers. All 4 feet are webbed. The long tail is thick at the base and gradually tapers to the tip. Males are larger than females.

OTTERS

HABITAT: Northern River Otters may visit coastal areas for brief periods, or they may live at the coast year-round. They sometimes roam far from water, especially young males establishing new territories. When roaming, a river otter rests under roots or overhangs, in hollow logs, in the abandoned burrows of other mammals or in ledges on rocky outcroppings.

FEEDING: On the coast, river otters feed primarily on fish such as sculpin, gunnel and flatfish. Crabs and shellfish are eaten when available. Coastal river otters also occasionally eat small terrestrial animals, such as mice, insects and earthworms.

REPRODUCTION: In coastal areas, the den is in a burrow or crevice away from the shore. The female bears a litter of 1 to 6 blind, fully furred young in March or April. The young are 5 ounces at birth. They first leave the den at 3 to 4 months, and they leave their parents at 6 to 7 months. River otters become sexually mature at 2 years. The mother breeds again soon after her litter is born, but delayed implantation of the embryos puts off the birth until the following spring.

SIMILAR SPECIES: The American Mink (*Neovison vison*) could be confused with a river otter, but although it sometimes lives in coastal marshes, it prefers freshwater habitats because its fur is not very tolerant of saltwater.

West Indian Manatee
Trichechus manatus

The unique West Indian Manatee is a favorite among wildlife enthusiasts. These gentle herbivores are found in shallow coastal waters where boaters can easily watch them as they feed on sea grass and bob up to the surface for a breath. The downside of their slow pace and preference for shallow waters is that they are at great risk to injuries and death from collisions with boats. It is not uncommon to see a manatee that bears the long, parallel scars of its encounter with an outboard motor. As awareness of their susceptibility has increased, a number of different policies and boat modifications have been implemented to protect the manatees. Boat speed limits are enforced in important manatee habitats, and some critical areas are fully protected as sanctuaries. Propeller guards help reduce injuries to manatees, but the majority of deaths are caused by hull collisions.

Interestingly, although manatees superficially resemble large seals, their closest relatives are actually the enormous elephant and diminutive hyrax. The Florida Manatee (*T. m. latirostris*) is one of two subspecies of the West Indian Manatee. The other subspecies is the Antillean Manatee (*T. m. manatus*) in Central America and the Caribbean. Recent genetical analysis indicates that there may be three subspecies, but at this time only two are recognized.

MANATEES

RANGE: found in coastal areas of the Atlantic from the Carolinas to Brazil, including the Caribbean.

OTHER NAMES: Florida Manatee, American Manatee.

STATUS: vulnerable; decreasing.

ADULT LENGTH: up to 15 ft; average 10 ft.

ADULT WEIGHT: up to 3300 lb; average 2200 lb.

BIRTH LENGTH: about 4 ft.

BIRTH WEIGHT: about 70 lb.

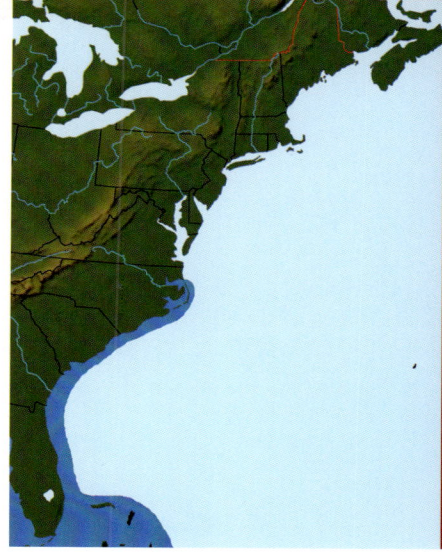

DESCRIPTION: Looking a bit like an overweight seal, the manatee has a massive, fusiform body. The rotund body tapers rapidly at the tail stock, and the tail is wide and spatula-shaped. The overall body color is light or brownish gray, often with a mottled appearance from algae that grows over its back. It has sparse, bristly hair over its entire body. The paddle-like flippers are curved and have 3 or 4 nails each. Its head is large and wrinkly, with small, dark eyes. Small ear openings are located directly behind the eyes. It has a large, whiskered muzzle with strong lips for grasping vegetation. The upper lip is split, and each side can move independently to grab and maneuver plant material into the mouth. The round nostrils are on the top of the nose, and they are pinched tightly shut while the animal is underwater. Females are slightly larger than males.

MANATEES

HABITAT: Although they tolerate fresh water easily, manatees are more common in shallow coastal waters. When they do enter fresh water, they mainly choose slow-moving rivers with abundant vegetation. Manatees are warm-water mammals, and though they have a layer of blubber, they cannot tolerate temperatures below 68° F. During the winter months they gather in warm Florida waters, especially near man-made heat sources such as power plants. Blue Spring, on Florida's east coast, has a relatively stable year-round temperature, and manatees gather here in large numbers during winter. During particularly warm summers they may be found as far north as Massachusetts, and there is one record of a manatee in waters off Rhode Island. Some Florida lakes have year-round populations.

FEEDING: Manatees are herbivores, and they rely on sea grass and other aquatic and marine plants. In Florida they have

MANATEES

REPRODUCTION: Manatees are generally solitary animals, but they are not territorial and easily accommodate the company of others, especially in warm wintering waters. Mating herds form, which are essentially a reproductive female pursued by several males. Mating occurs in any season but peaks in summer. After a gestation of 11 to 14 months, a single calf is born. Twins are possible, but rare. A calf is nursed for about 1 year and stays with its mother for 2 years. Like an elephant, the female has 2 small teats located near the forelimbs. The bond between a mother and her calf is very strong, and a calf rarely strays more than a few feet away. Touching seems to be extremely important between the mother and calf, and the two also have a variety of specialized clicks and chirps for communication. Females are sexually mature when they are 4 or 5 years old, and males when they are 3 or 4 years old. Lifespan is well over 30 years, and there is a record of a captive female giving birth in her mid-40s.

SIMILAR SPECIES: A manatee is unlikely to be confused with any other mammal in the region.

a preference for grasses in salt marshes, so they time their feeding periods with high tides. Sea grasses are relatively low in nutrients, and manatees need to graze for 6 to 8 hours a day. On average, a manatee consumes from 5 to 10 percent of its body weight in plant material each day, which for a large individual can be more than 200 pounds of food. Their molars wear down quickly, and new molars constantly erupt behind the old ones. The old molars are pushed forward as they wear out.

Glossary

AMPHIPOD: a shrimplike crustacean (e.g., krill) of the order Amphipoda that has a laterally compressed body; the primary food for many whale species

BALEEN: strands of keratin that hang in sheets from the upper jaw of whales in the suborder Mysticeta and are used to filter food from water

BENTHIC: dwelling at or near the bottom of a lake or ocean

BOW-RIDING: a behavior seen in dolphins in which the animals swim in the bow waves of boats

BREACH: a whale display in which the animal rises vertically into the air, clearing the water's surface with almost all of its body before splashing back in

BULL: an adult male cetacean or pinniped

CALF: a baby cetacean

CALLOSITY: an area of skin that is hardened by growths of lumpy, keratinous material; commonly found on the head of a right whale

CETACEAN: a marine mammal of the order Cetacea, which encompasses all whale, dolphin and porpoise species

COLONY: a group of animals living together and interacting socially

COPEPOD: any of a large group of small marine and aquatic crustaceans, including amphipods; the primary food for many marine creatures

COW: an adult female cetacean or pinniped

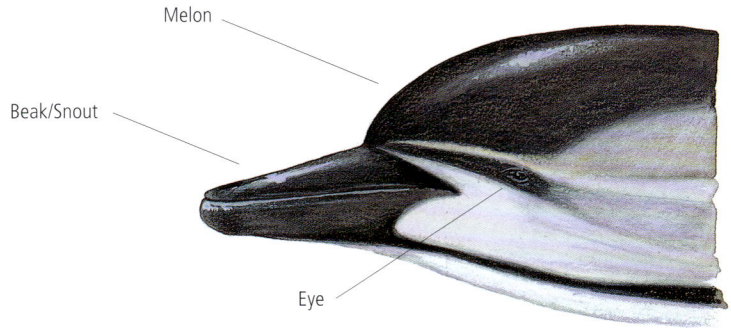

GLOSSARY

DELPHINID: a member of the ocean dolphin family (Delphinidae)

DORSAL: pertaining to the back or spine (compare **VENTRAL**)

DORSUM: the back or dorsal region of the body

ECHOLOCATION: the detection of an object by emitting sound waves and interpreting the returning echoes, which are changed by bouncing off the object

ENCEPHALIZATION QUOTIENT (EQ): the ratio of an animal's brain weight to its body size; used as a measurement of the possible intelligence of an animal

ENDANGERED: facing imminent extirpation or extinction

EXTINCT: no longer in existence anywhere

EXTIRPATED: no longer found in a given geographic area but still in existence elsewhere in the world

FALCATE: curved or somewhat sickle-shaped

FLUKE: *n.* either of the lobes on a whale's tail; *v.* a whale behavior in which the animal raises its tail above the water before diving

FUSIFORM: spindle-shaped; having an elongated body that tapers at both ends

GESTATION: the time of pregnancy, from conception to birth

KERATINOUS: referring to the material that composes hair, wool and nails in many mammals

KRILL: a general name to describe more than 80 species of small, shrimplike organisms that are eaten by numerous marine creatures

HABITAT: the environment in which an animal or plant lives

HOME RANGE: the total area through which an individual animal moves during its usual activities (compare **TERRITORY**)

LANUGO: the fine, soft hair that covers the body of newborn mammals

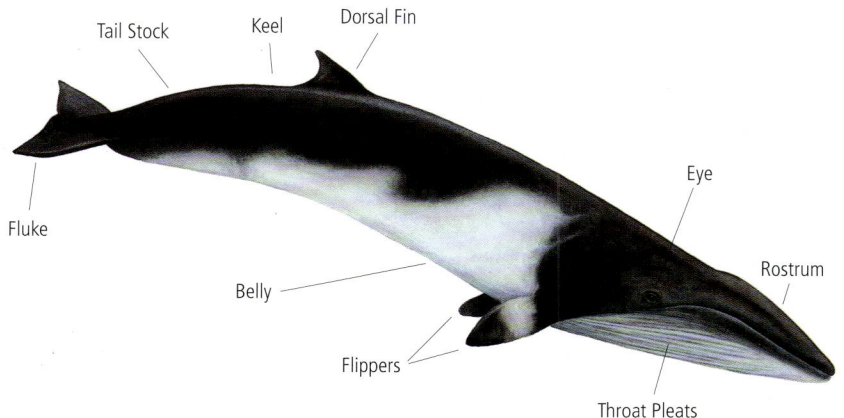

GLOSSARY

LARYNGEAL: referring to the larynx in the region of the throat, which can be made to vibrate to produce sound

LOB-TAIL: a display in which a whale forcefully slaps its tail flukes on the water's surface

MELON: an ovoid-shaped fatty organ found in the forehead of all toothed whales and believed to be used in echolocation

MIGRATION: the journey that an animal undertakes to get from one region to another, usually in response to seasonal and reproductive cycles

MYSTICETE: any cetacean that has baleen instead of teeth

NOMADIC: referring to creatures that move from region to region solely in response to food availability and that otherwise do not migrate

ODONTOCETE: any cetacean that has teeth instead of baleen

PINNIPED: a carnivorous, aquatic mammal of the suborder Pinnipedia (order Carnivora) with a streamlined body specialized for swimming and limbs modified into flippers (e.g., seals, sea lions, walruses)

PLANKTON: the tiny plants (phytoplankton) and animals (zooplankton) that drift in the water column and are the base of the food chain

POD: a small herd or school of whales

PORPOISE: *v.* a method of moving through the water that involves alternately rising above the water's surface and then diving underwater

PUP: a baby pinniped

ROOKERY: a colony of breeding animals

RORQUAL: strictly used, a whale in the genus *Balaenoptera*; most authorities also include the Humpback Whale

ROSTRUM: the forward-projecting snout of a cetacean

SONAR: the ability to detect or track objects underwater by monitoring their sound emissions or how sound bounces off them

SPY-HOP: a whale behavior in which the animal, while in an almost vertical position, raises its head out of the water just far enough to look around

STRANDING: an event where one or more cetaceans come onto land or into shallow water and become stuck

TERRITORY: a defended area within an animal's home range

THREATENED: likely to become endangered if limiting factors are not reversed

TUBERCLE: a small, rounded nodule on an animal's skin

VENTRAL: pertaining to the belly (compare **DORSAL**)

VESTIGIAL: a part or organ of a creature that has atrophied or is rendered functionless by the process of evolution

VULNERABLE: refers to a species that is of special concern because of characteristics that make it particularly sensitive to human activities or natural events

Further Information

Emergency Hotlines on the East Coast

Northeast Region Marine Mammal and Sea Turtle Stranding & Entanglement Hotline
866-755-6622

NOAA Fisheries Stranding Hotline
978-281-9351

NMFS Southeast Marine Mammal Stranding Hotline
877-433-8299

Local Contacts and Research & Conservation Groups

Northeast Region

- *Maine*

Maine Marine Animal Reporting Hotline
800-532-9551

Maine Department of Marine Resources
Boothbay Harbor
207-633-9500
www.maine.gov/dmr

Allied Whale, College of the Atlantic
Bar Harbor
207-288-5644
www.coa.edu/allied-whale-microsite.htm

- *Massachusetts*

Protected Species Branch, NMFS Northeast Fisheries Science Center
Woods Hole
508-495-2000
www.nefsc.noaa.gov

FURTHER INFORMATION

Protected Resources Division, NMFS Northeast Region
Gloucester
978-281-9300
www.nefsc.noaa.gov

Whale Center of New England
Gloucester
978-281-6351
www.whalecenter.org

National Park Service, Cape Cod National Seashore
Wellfleet
508-349-3785
www.nps.gov/caco

New England Aquarium
Boston
617-973-5247
www.neaq.org

International Fund for Animal Welfare, Marine Mammal Rescue Program (formerly known as the Cape Cod Stranding Network)
Yarmouth Port
800-932-4329
508-744-2000
www.ifaw.org

National Marine Life Center
Buzzards Bay
508-743-9888
nmlc.org

- *Connecticut/Rhode Island*

Mystic Aquarium
Mystic, CT
860-572-5955 x107
www.mysticaquarium.org

- *New York*

Riverhead Foundation for Marine Research and Preservation
Riverhead
24-hour Rescue Hotline: 631-369-9829
www.riverheadfoundation.org

FURTHER INFORMATION

- *New Jersey*

Marine Mammal Stranding Center
Brigantine
609-266-0538
www.marinemammalstrandingcenter.org

- *Delaware*

MERR Institute
Nassau
302-228-5029
merrinstitute.org

- *Maryland*

National Aquarium in Baltimore, Marine Animal Rescue Program
Baltimore
410-576-1098
www.aqua.org

**Maryland Department of Natural Resources
Cooperative Oxford Laboratory**
Oxford
800-628-9944
www.dnr.state.md.us

Southeast Region

- *Virginia*

Virginia Aquarium & Marine Science Museum
Virginia Beach
757-385-3474
www.virginiaaquarium.com

Virginia Institute of Marine Science, College of William and Mary
Gloucester Point
804-684-7000
www.vims.edu

Smithsonian Institute, National Museum of Natural History
Washington, DC
202-633-1260
www.si.edu

FURTHER INFORMATION

- *North Carolina*

North Carolina Aquariums
Raleigh
800-832-3474

Cape Hatteras National Seashore
Manteo
252-473-2595
www.nps.gov/caha/index.htm

North Carolina Wildlife Resources Commission
800-662-7137
919-707-0040
www.ncwildlife.org

Duke University Marine Laboratory
Beaufort
252-504-7503
www.nicholas.duke.edu/marinelab

- *South Carolina*

Stranding Hotline, Natural Resources Department
800-922-5431

- *Georgia*

Stranding Hotline, Georgia Natural Resources Department
Brunswick
912-264-7218

- *Florida*

Marine Animal Rescue Society/Stranding Hotline
Miami
888-404-3922
www.marineanimalrescue.org

Dolphin Conservation Field Station at Marineland
St. Augustine
877-933-3402
www.marineland.net/dcfs.php

Florida Aquarium
Tampa
813-273-4000
www.flaquarium.org/

Florida Fish and Wildlife Conservation Commission
myfwc.com

FURTHER INFORMATION

Hubbs-SeaWorld Research Institute
321-327-8970
www.hswri.org

Marine Mammal Conservancy
Key Largo
marinemammalconservancy.org

Online Information

American Cetacean Society
www.acsonline.org

Cetacea
www.cetacea.org

Convention on International Trade of Endangered Species of Wild Flora and Fauna (CITES)
www.cites.org

Dolphin Research Center
www.dolphins.org

Dolphin Institute
www.dolphin-institute.com

International Marine Mammal Project
www.earthisland.org/immp/index.html

Ocean Link
oceanlink.island.net/oinfo/oinfo.html

Project Aware (Aquatic World Awareness, Responsibility and Education)
www.projectaware.org

Seal Conservation Society
www.pinnipeds.org/main.htm

Shedd Aquarium
www.sheddaquarium.org

Whale and Dolphin Conservation Society
www.wdcs-na.org/

Whalenet
whale.wheelock.edu/Welcome.html

Whales on the Net
www.whales.org.au

FURTHER INFORMATION

Further Reading

Boschung, H.T., Jr., J.D. Williams, D.W. Gotshall, D.K Caldwell and M.C. Caldwell. 1983. *The Audubon Society Field Guide to North American Fishes, Whales and Dolphins*. Alfred A. Knopf, New York.

Bulloch, D.K. 1993. *The Whale-watcher's Handbook: A Guide to the Whales, Dolphins and Porpoises of North America*. Lyons & Burford, New York.

Carwardine, M. 1995. *Whales, Dolphins and Porpoises*. Dorling Kindersley, London.

Carwardine, M., E. Hoyt, R.E. Fordyce and P. Gill. 1998. *Whales, Dolphins & Porpoises*. Nature Company: Time-Life Books, Alexandria, Virginia.

Clapham, P. 1997. *Whales of the World*. Voyageur Press, Stillwater, Minnesota.

Conner, R.C., and D.M. Peterson. 1994. *The Lives of Whales and Dolphins*. Henry Holt and Company, New York.

Corrigan, P. 1991. *Where the Whales Are: Your Guide to Whale-watching Trips in North America*. The Globe Pequot Press, Chester, Connecticut.

Darling, J.D., C. Nicklin, K.S. Norris, H. Whitehead and B. Würsig. 1995. *Whales, Dolphins and Porpoises*. National Geographic Society, Washington, D.C.

Gordon, D.G., and C. Flaherty. 1990. *The American Cetacean Society Field Guide to the Orca*. Sasquatch Books, Seattle.

Kraus, S.D., and R.M. Rolland. 2007. *The Urban Whale: North Atlantic Right Whales at the Crossroads*. President and Fellows of Harvard College. USA.

Mason, A. 1999. *Whales, Dolphins & Porpoises*. Altitude Publishing, Canmore, Alberta.

Reader's Digest. 1997. *Whales, Dolphins and Porpoises*. Reader's Digest Explores Science & Nature Series. The Reader's Digest Association, Pleasantville, New York.

Rollins D., and E. Byrd. 1995. "The Hello Dolphin Project." First International Symposium on Dolphin Assisted Therapy, Sept 8–10, Cancun Mexico, Proceedings: 17.

Stewart, F., ed. 1995. *The Presence of Whales: Contemporary Writings on the Whale*. Alaska Northwest Books, Anchorage.

Stonehouse, B. 1998. A *Visual Introduction to Whales, Dolphins and Porpoises*. Checkmark Books, New York.

Index

Page numbers in **boldface** type refer to the primary species account.

A
Atlantic Spotted Dolphin. *See* Dolphin, Atlantic Spotted
Atlantic White-sided Dolphin. *See* Dolphin, Atlantic White-sided

B
Balaenidae, **47**, 66–69
Balaenoptera
    acutorostrata, **48–51**
    borealis, **52–53**
    edeni, **54–55**
    musculus, **56–59**
    physalus, **60–61**
Balaenopteridae, **46**, 48–65
Beaked Whale Family, **70–71**, 118–29
Blackfish. *See* Whale, False Killer. *See also* Whale, Long-finned Pilot. *See also* Whale, Short-finned Pilot
    Many-toothed. *See* Whale, Melon-headed
    Slender. *See* Whale, Pygmy Killer
Blainville's Beaked Whale. *See* Whale, Blainville's Beaked
Blue Whale. *See* Whale, Blue
Bottlehead. *See* Whale, North Atlantic Bottlenose
Bottlenose Dolphin. *See* Dolphin, Bottlenose
Bryde's Whale. *See* Whale, Bryde's

C
Cachalot. *See* Whale, Sperm
    Lesser. *See* Whale, Pygmy Sperm
Clymene Dolphin. *See* Dolphin, Clymene
Cowfish. *See* Dolphin, Bottlenose
Cuvier's Beaked Whale. *See* Whale, Cuvier's Beaked

D
Delphinidae, **70**, 72–113

Delphinus delphis, **88–91**
Dolphin
    Atlantic Bottlenose. *See* Bottlenose D.
    Atlantic Spinner. *See* Clymene D.
    Atlantic Spotted, 77, 79, 83, **84–85**
    Atlantic White-sided, 91, **92–95**, 97, 117
    Bottle-nosed. *See* Bottlenose D.
    Bottlenose, 73, **74–77**, 79, 85, 87, 99
    Bridled. *See* Pantropical Spotted D.
    Clymene, 77, **80–81**, 83, 87
    Common. *See* Short-beaked Saddleback D.
    Criss-cross. *See* Short-beaked Saddleback D.
    Cuvier's. *See* Atlantic Spotted D.
    Electra. *See* Whale, Melon-headed
    Gray. *See* Bottlenose D.
    Gulf Stream Spotted. *See* Atlantic Spotted D.
    Helmet. *See* Clymene D.
    Long-snouted. *See* Atlantic Spotted D. *See also* Spinner D.
    Long-snouted Spinner. *See* Spinner D.
    Pantropical Spotted, 73, **78–79**, 83, 85
    Risso's, 77, **98–99**, 101, 103, 131, 133
    Rough-toothed, **72–73**, 77, 87
    Short-beaked Common. *See* Short-beaked Saddleback D.
    Short-beaked Saddleback, **88–91**, 95, 97
    Short-snouted Spinner. *See* Clymene D.
    Slender-beaked. *See* Pantropical Spotted D.
    Spinner, 73, 77, 79, 81, 83, 85, **86–87**
    Spotted. *See* Pantropical Spotted D.
    Striped, 81, **82–83**, 91, 95, 97
    White-beaked, 91, 95, **96–97**, 117
    White-spotted. *See* Pantropical Spotted D.

INDEX

Dwarf Sperm Whale. *See* Whale, Dwarf Sperm
Dwarf Sperm Whale Family, **71**, **130–33**

E
Eubalaena glacialis, **66–69**

F
False Killer Whale. *See* Whale, False Killer
Feresa attenuata, **102–03**
Finback. *See* Whale, Fin
    Lesser. *See* Whale, Northern Minke
Finner. *See* Whale, Fin
    Japan. *See* Whale, Sei
    Little. *See* Whale, Northern Minke
    Sharp-headed. *See* Whale, Northern Minke
Fin Whale. *See* Whale, Fin

G
Gervais's Beaked Whale. *See* Whale, Gervais's Beaked
Globicephala
    macrorhynchus, **106–07**
    melas, **108–09**
Grampus griseus, **98–99**
Gray Seal. *See* Seal, Gray

H
Hair Seal Family, **138**, **140–47**
Halichoerus grypus, **144–45**
Harbor Porpoise. *See* Porpoise, Harbor
Harbor Seal. *See* Seal, Harbor
Harp Seal. *See* Seal, Harp
Humpback Whale. *See* Whale, Humpback
Hyperoodon ampullatus, **120–21**

K
Kogia
    breviceps, **130–31**
    sima, **132–33**
Kogiidae, **71**, **130–33**

L
Lagenorhynchus
    acutus, **92–95**
    albirostris, **96–97**
Long-finned Pilot Whale. *See* Whale, Long-finned Pilot
Lontra canadensis, **148–51**
Lutrinae, **139**, **148–51**

M
Manatee
    American. *See* West Indian M.
    Florida. *See* West Indian M.
    West Indian, **152–55**
Manatee Family, **139**, **152–55**
Megaptera novaeangliae, **62–65**
Melon-headed Whale. *See* Whale, Melon-headed

Mesoplodon
    bidens, **122–23**
    densirostris, **124–25**
    europaeus, **126–27**
    mirus, **128–29**

N
North Atlantic Bottlenose Whale. *See* Whale, North Atlantic Bottlenose
North Atlantic Right Whale. *See* Whale, North Atlantic Right
Northern Minke Whale. *See* Whale, Northern Minke
Northern River Otter. *See* Otter, Northern River

O
Ocean Dolphin Family, **70**, **72–113**
Orca. *See* Whale, Killer
Orcinus orca, **110–13**
Otter, Northern River, **148–51**
Otter Subfamily, **139**, **148–51**

P
Pantropical Spotted Dolphin. *See* Dolphin, Pantropical Spotted
Peponocephala electra, **100–01**
Phoca
    groenlandica, **146–47**
    vitulina, **140–43**
Phocidae, **138**, **140–47**
Phocoena phocoena, **114–17**
Phocoenidae, **70**, **114–17**
Physeteridae, **71**, **134–37**
Physeter macrocephalus, **134–37**
Porpoise
    Black. *See* Dolphin, Rough-toothed
    Common. *See* Harbor P.
    Gray. *See* Dolphin, Bottlenose
    Harbor, **114–17**
    Long-beaked. *See* Dolphin, Spinner
    Puffing. *See* Harbor P.
    Rough-toothed. *See* Dolphin, Rough-toothed
Porpoise Family, **70**, **114–17**
Pseudorca. *See* Whale, False Killer
Pseudorca crassidens, **104–05**
Pygmy Killer Whale. *See* Whale, Pygmy Killer
Pygmy Sperm Whale. *See* Whale, Pygmy Sperm

R
Razorback. *See* Whale, Fin
Right Whale Family, **47**, **66–69**
Risso's Dolphin. *See* Dolphin, Risso's
Rollover. *See* Dolphin, Spinner
Rorqual
    Boreal. *See* Whale, Sei
    Common. *See* Whale, Fin
    Great Northern. *See* Whale, Blue

INDEX

Lesser. *See* Whale, Northern Minke
Rudolphi's. *See* Whale, Sei
Sibbald's. *See* Whale, Blue
Rorqual Family, **46**, 48–65
Rough-toothed Dolphin. *See* Dolphin, Rough-toothed

S

Sea Cow Family. *See* Manatee Family
Seal
    Atlantic. *See* Gray S.
    Common. *See* Harbor S.
    Gray, 143, **144–45**, 147
    Greenland. *See* Harp S.
    Harbor, **140–43**, 145, 147
    Harp, 143, 145, **146–47**
    Horsehead. *See* Gray S.
    Jumping. *See* Harp S.
    Saddle-backed. *See* Harp S.
Sei Whale. *See* Whale, Sei
Short-beaked Saddleback Dolphin. *See* Dolphin, Short-beaked Saddleback
Short-finned Pilot Whale. *See* Whale, Short-finned Pilot
Slopehead. *See* Dolphin, Rough-toothed
Sowerby's Beaked Whale. *See* Whale, Sowerby's Beaked
Sperm Whale. *See* Whale, Sperm
Sperm Whale Family, **71**, 134–37
Spinner Dolphin. *See* Dolphin, Spinner
Stenella
    attenuata, **78–79**
    clymene, **80–81**
    coeruleoalba, **82–83**
    frontalis, **84–85**
    longirostris, **86–87**
Steno. *See* Dolphin, Rough-toothed
Steno bredanensis, **72–73**
Striped Dolphin. *See* Dolphin, Striped
Sulfur-bottom. *See* Whale, Blue

T

Trichechidae, **139**, 152–55
Trichechus manatus, **152–55**
True's Beaked Whale. *See* Whale, True's Beaked
Tursiops truncatus, **74–77**

W

West Indian Manatee. *See* Manatee, West Indian
Whale
    Antillean Beaked. *See* Gervais's Beaked W.
    Anvil-headed. *See* Sperm W.
    Atlantic Beaked. *See* Blainville's Beaked W.
    Biscayan Right. *See* North Atlantic Right W.
    Black Right. *See* North Atlantic Right W.
    Blainville's Beaked, **124–25**

    Blue, 53, 55, **56–59**, 61
    Bryde's, 51, 53, **54–55**, 59, 61
    Coalfish. *See* Sei W.
    Common Minke. *See* Northern Minke W.
    Cuvier's. *See* Cuvier's Beaked W.
    Cuvier's Beaked, **118–19**, 123, 127, 129
    Dense Beaked. *See* Blainville's Beaked W.
    Dwarf Sperm, 99, 131, **132–33**
    European Beaked. *See* Gervais's Beaked W.
    False Killer, 101, 103, **104–05**, 107, 109, 113
    False Pilot. *See* False Killer W.
    Fin, 53, 59, **60–61**
    Gervais's Beaked, 119, 123, **126–27**, 129
    Goose-beaked. *See* Cuvier's Beaked W.
    Gulf Stream Beaked. *See* Gervais's Beaked W.
    Herring. *See* Fin W.
    Humpback, **62–65**, 69, 137
    Killer, 105, 107, 109, **110–13**
    Lesser Sperm. *See* Pygmy Sperm W.
    Little Killer. *See* Melon-headed W.
    Long-finned Pilot, 107, **108–09**, 113
    Melon-headed, 99, **100–01**, 103, 105, 109
    North Atlantic Beaked. *See* Sowerby's Beaked W.
    North Atlantic Bottlenose, **120–21**
    North Atlantic Right, 65, **66–69**
    Northern Bottlenose. *See* North Atlantic Bottlenose W.
    Northern Minke, **48–51**, 53, 55
    Northern Right. *See* North Atlantic Right W.
    North Sea Beaked. *See* Sowerby's Beaked W.
    Pacific Pilot. *See* Short-finned Pilot W.
    Piked. *See* Northern Minke W.
    Pollack. *See* Sei W.
    Pothead. *See* Long-finned Pilot W. *See also* Short-finned Pilot W.
    Pygmy Killer, 101, **102–03**, 105, 109
    Pygmy Sperm, 99, **130–31**, 133
    Sardine. *See* Sei W.
    Sei, 51, **52–53**, 59, 61
    Short-finned Pilot, 105, **106–07**, 109, 113
    Short-headed. *See* Pygmy Sperm W.
    Slender Pilot. *See* Pygmy Killer W.
    Sowerby's Beaked, 119, **122–23**, 127, 129
    Sperm, **134–37**
    Tropical. *See* Bryde's W.
    Tropical Beaked. *See* Blainville's Beaked W.
    True's Beaked, 119, 123, 127, **128–29**
    Wonderful Beaked. *See* True's Beaked W.
    White-beaked Dolphin. *See* Dolphin, White-beaked

Z

Ziphiidae, **70–71**, 118–29
Ziphius cavirostris, **118–19**

About the Author

TAMARA EDER, equipped from the age of six with a canoe, a dip net and a notepad, grew up with a fascination for nature and the diversity of life. She has a degree in environmental conservation sciences, and has photographed and written about wildlife in Bermuda, the Galapagos Islands, the Amazon Basin, Argentina, Tibet and India. With a fondness for paleontology, Tamara has studied and participated in paleo digs in Alberta and Patagonia. She has worked both as an interpretive naturalist and guide, specializing in ecology and paleontology. An award-winning photographer, her photographs appear in numerous books, posters and online magazines. Tamara now lives in Patagonia, and she continues to write, photograph and travel.

About the Illustrator

IAN SHELDON has been captivated by creatures large and small since the age of three. Born in Canada, Ian later lived in South Africa, England and Singapore. Exposure to nature from so many different places enhanced his desire to study it further, and he earned an award from the Zoological Society of London and a degree from Cambridge University. He has also completed a Master's degree in Ecotourism Development. Ian is an accomplished artist whose diverse works have been shown in San Francisco, New York, across Canada and in England. He has illustrated or authored over 30 books, many of them bestsellers.